Property of
The Public Library of Nashville and Davidson County
225 Polk Ave., Nashville, TN 37203

SOLAR ENERGY

EARTH • AT • RISK

Acid Rain
Alternative Sources of Energy
Animal Welfare
The Automobile and the Environment
Clean Air
Clean Water
Degradation of the Land
Economics and the Environment
Environmental Action Groups
Environmental Disasters
The Environment and the Law
Extinction
The Fragile Earth
Global Warming
The Living Ocean
Nuclear Energy • Nuclear Waste
Overpopulation
The Ozone Layer
The Rainforest
Recycling
Solar Energy
Toxic Materials
What You Can Do for the Environment
Wilderness Preservation

EARTH • AT • RISK

SOLAR ENERGY

by Bob Brooke

Introduction by
Russell E. Train

Chairman of
the Board of Directors,
World Wildlife Fund and
The Conservation Foundation

CHELSEA HOUSE PUBLISHERS
new york   philadelphia

CHELSEA HOUSE PUBLISHERS
EDITOR-IN-CHIEF: Remmel Nunn
MANAGING EDITOR: Karyn Gullen Browne
COPY CHIEF: Mark Rifkin
PICTURE EDITOR: Adrian G. Allen
ART DIRECTOR: Maria Epes
ASSISTANT ART DIRECTOR: Howard Brotman
MANUFACTURING DIRECTOR: Gerald Levine
SYSTEMS MANAGER: Lindsey Ottman
PRODUCTION MANAGER: Joseph Romano
PRODUCTION COORDINATOR: Marie Claire Cebrián

EARTH AT RISK
SENIOR EDITOR: Jake Goldberg

Staff for *Solar Energy*
ASSOCIATE EDITOR: Karen Hammonds
SENIOR COPY EDITOR: Laurie Kahn
EDITORIAL ASSISTANT: Danielle Janusz
PICTURE RESEARCHER: Pat Burns
SERIES DESIGNER: Maria Epes
SENIOR DESIGNER: Marjorie Zaum
COVER ILLUSTRATION: Shelley Pritchett

Copyright © 1992 by Chelsea House Publishers, a division of Main Line Book Co. All rights reserved. Printed and bound in the United States of America.

5 7 9 8 6 4

Library of Congress Cataloging-in-Publication Data
Brooke, Bob
 Solar energy/by Bob Brooke; introduction by Russell E. Train.
   p. cm.—(Earth at risk)
 Includes bibliographical references and index.
 Summary: Discusses the history, uses, advantages, and disadvantages of solar energy and examines recent advances in that field.
 ISBN 0-7910-1590-4
       0-7910-1615-3 (pbk.)
 1. Solar energy—Juvenile literature. [1. Solar energy.] I. Title. II. Series.   91-24298
TJ810.3.B76 1992   CIP
333.792'3—dc20   AC

CONTENTS

Introduction—Russell E. Train   6

1 The Ultimate Energy Source   13

2 Early Uses of Solar Energy   23

3 Collecting Sunshine   37

4 Photovoltaic Cells   49

5 The Solar House   59

6 Solar Energy Today   71

7 The Pros and Cons of Solar Energy   81

8 The Solar Future   89

   Appendix: For More Information   99

   Conversion Table   100

   Further Reading   101

   Glossary   103

   Index   107

INTRODUCTION

Russell E. Train
Administrator, Environmental Protection Agency, 1973 to 1977; Chairman of the Board of Directors, World Wildlife Fund and The Conservation Foundation

There is a growing realization that human activities increasingly are threatening the health of the natural systems that make life possible on this planet. Humankind has the power to alter nature fundamentally, perhaps irreversibly.

This stark reality was dramatized in January 1989 when *Time* magazine named Earth the "Planet of the Year." In the same year, the Exxon *Valdez* disaster sparked public concern over the effects of human activity on vulnerable ecosystems when a thick blanket of crude oil coated the shores and wildlife of Prince William Sound in Alaska. And, no doubt, the 20th anniversary celebration of Earth Day in April 1990 renewed broad public interest in environmental issues still further. It is no accident then that many people are calling the years between 1990 and 2000 the "Decade of the Environment."

And this is not merely a case of media hype, for the 1990s will truly be a time when the people of the planet Earth learn the meaning of the phrase "everything is connected to everything else" in the natural and man-made systems that sustain our lives. This will be a period when more people will understand that burning a tree in Amazonia adversely affects the global atmosphere just as much as the exhaust from the cars that fill our streets and expressways.

Central to our understanding of environmental issues is the need to recognize the complexity of the problems we face and the relationships between environmental and other needs in our society. Global warming provides an instructive example. Controlling emissions of carbon dioxide, the principal greenhouse gas, will involve efforts to reduce the use of fossil fuels to generate electricity. Such a reduction will include energy conservation and the promotion of alternative energy sources, such as nuclear and solar power.

The automobile contributes significantly to the problem. We have the choice of switching to more energy efficient autos and, in the longer run, of choosing alternative automotive power systems and relying more on mass transit. This will require different patterns of land use and development, patterns that are less transportation and energy intensive.

In agriculture, rice paddies and cattle are major sources of greenhouse gases. Recent experiments suggest that universally used nitrogen fertilizers may inhibit the ability of natural soil organisms to take up methane, thus contributing tremendously to the atmospheric loading of that gas—one of the major culprits in the global warming scenario.

As one explores the various parameters of today's pressing environmental challenges, it is possible to identify some areas where we have made some progress. We have taken important steps to control gross pollution over the past two decades. What I find particularly encouraging is the growing environmental consciousness and activism by today's youth. In many communities across the country, young people are working together to take their environmental awareness out of the classroom and apply it to everyday problems. Successful recycling and tree-planting projects have been launched as a result of these budding environmentalists who have committed themselves to a cleaner environment. Citizen action, activated by youthful enthusiasm, was largely responsible for the fast-food industry's switch from rainforest to domestic beef, for pledges from important companies in the tuna industry to use fishing techniques that would not harm dolphins, and for the recent announcement by the McDonald's Corporation to phase out polystyrene "clam shell" hamburger containers.

Despite these successes, much remains to be done if we are to make ours a truly healthy environment. Even a short list of persistent issues includes problems such as acid rain, ground-level ozone and smog, and airborne toxins; groundwater protection and nonpoint sources of pollution, such as runoff from farms and city streets; wetlands protection; hazardous waste dumps; and solid waste disposal, waste minimization, and recycling.

Similarly, there is an unfinished agenda in the natural resources area: effective implementation of newly adopted management plans for national forests; strengthening the wildlife refuge system; national park management, including addressing the growing pressure of development on lands surrounding the parks; implementation of the Endangered Species Act; wildlife trade problems, such as that involving elephant ivory; and ensuring adequate sustained funding for these efforts at all levels of government. All of these issues are before us today; most will continue in one form or another through the year 2000.

Each of these challenges to environmental quality and our health requires a response that recognizes the complex nature of the problem. Narrowly conceived solutions will not achieve lasting results. Often it seems that when we grab hold of one part of the environmental balloon, an unsightly and threatening bulge appears somewhere else.

The higher environmental issues arise on the national agenda, the more important it is that we are armed with the best possible knowledge of the economic costs of undertaking particular environmental programs and the costs associated with not undertaking them. Our society is not blessed with unlimited resources, and tough choices are going to have to be made. These should be informed choices.

All too often, environmental objectives are seen as at cross-purposes with other considerations vital to our society. Thus, environmental protection is often viewed as being in conflict with economic growth, with energy needs, with agricultural productions, and so on. The time has come when environmental considerations must be fully integrated into every nation's priorities.

One area that merits full legislative attention is energy efficiency. The United States is one of the least energy efficient of all the industrialized nations. Japan, for example, uses far less energy per unit of gross national product than the United States does. Of course, a country as large as the United States requires large amounts of energy for transportation. However, there is still a substantial amount of excess energy used, and this excess constitutes waste. More fuel efficient autos and home heating systems would save millions of barrels of oil, or their equivalent, each year. And air pollutants, including greenhouse gases, could be significantly reduced by increased efficiency in industry.

I suspect that the environmental problem that comes closest to home for most of us is the problem of what to do with trash. All over the world, communities are wrestling with the problem of waste disposal. Landfill sites are rapidly filling to capacity. No one wants a trash and garbage dump near home. As William Ruckelshaus, former EPA administrator and now in the waste management business, puts it, "Everyone wants you to pick up the garbage and no one wants you to put it down!"

At the present time, solid waste programs emphasize the regulation of disposal, setting standards for landfills, and so forth. In the decade ahead, we must shift our emphasis from regulating waste disposal to an overall reduction in its volume. We must look at the entire waste stream, including product design and packaging. We must avoid creating waste in the first place. To the greatest extent possible, we should then recycle any waste that is produced. I believe that, while most of us enjoy our comfortable way of life and have no desire to change things, we also know in our hearts that our "disposable society" has allowed us to become pretty soft.

Land use is another domestic issue that might well attract legislative attention by the year 2000. All across the United States, communities are grappling with the problem of growth. All too often, growth imposes high costs on the environment—the pollution of aquifers; the destruction of wetlands; the crowding of shorelines; the loss of wildlife habitat; and the loss of those special places, such as a historic structure or area, that give a community a sense of identity. It is worth noting that growth is not only the product of economic development but of population movement. By the year 2010, for example, experts predict that 75% of all Americans will live within 50 miles of a coast.

It is important to keep in mind that we are all made vulnerable by environmental problems that cross international borders. Of course, the most critical global conservation problems are the destruction of tropical forests and the consequent loss of their biological capital. Some scientists have calculated extinction rates as high as 11 species per hour. All agree that the loss of species has never been greater than at the present time; not even the disappearance of the dinosaurs can compare to today's rate of extinction.

In addition to species extinctions, the loss of tropical forests may represent as much as 20% of the total carbon dioxide loadings to the atmosphere. Clearly, any international approach to the problem of global warming must include major efforts to stop the destruction of forests and to manage those that remain on a renewable basis. Debt for nature swaps, which the World Wildlife Fund has pioneered in Costa Rica, Ecuador, Madagascar, and the Philippines, provide a useful mechanism for promoting such conservation objectives.

Global environmental issues inevitably will become the principal focus in international relations. But the single overriding issue facing the world community today is how to achieve a sustainable balance between growing human populations and the earth's natural systems. If you travel as frequently as I do in the developing countries of Latin America, Africa, and Asia, it is hard to escape the reality that expanding human populations are seriously weakening the earth's resource base. Rampant deforestation, eroding soils, spreading deserts, loss of biological diversity, the destruction of fisheries, and polluted and degraded urban environments threaten to spread environmental impoverishment, particularly in the tropics, where human population growth is greatest.

It is important to recognize that environmental degradation and human poverty are closely linked. Impoverished people desperate for land on which to grow crops or graze cattle are destroying forests and overgrazing even more marginal land. These people become trapped in a vicious downward spiral. They have little choice but to continue to overexploit the weakened resources available to them. Continued abuse of these lands only diminishes their productivity. Throughout the developing world, alarming amounts of land rendered useless by overgrazing and poor agricultural practices have become virtual wastelands, yet human numbers continue to multiply in these areas.

From Bangladesh to Haiti, we are confronted with an increasing number of ecological basket cases. In the Philippines, a traditional focus of U.S. interest, environmental devastation is widespread as deforestation, soil erosion, and the destruction of coral reefs and fisheries combine with the highest population growth rate in Southeast Asia.

Controlling human population growth is the key factor in the environmental equation. World population is expected to at least double to about 11 billion before leveling off. Most of this growth will occur in the poorest nations of the developing world. I would hope that the United States will once again become a strong advocate of international efforts to promote family planning. Bringing human populations into a sustainable balance with their natural resource base must be a vital objective of U.S. foreign policy.

Foreign economic assistance, the program of the Agency for International Development (AID), can become a potentially powerful tool for arresting environmental deterioration in developing countries. People who profess to care about global environmental problems—the loss of biological diversity, the destruction of tropical forests, the greenhouse effect, the impoverishment of the marine environment, and so on—should be strong supporters of foreign aid planning and the principles of sustainable development urged by the World Commission on Environment and Development, the "Brundtland Commission."

If sustainability is to be the underlying element of overseas assistance programs, so too must it be a guiding principle in people's practices at home. Too often we think of sustainable development only in terms of the resources of other countries. We have much that we can and should be doing to promote long-term sustainability in our own resource management. The conflict over our own rainforests, the old growth forests of the Pacific Northwest, illustrates this point.

The decade ahead will be a time of great activity on the environmental front, both globally and domestically. I sincerely believe we will be tested as we have been only in times of war and during the Great Depression. We must set goals for the year 2000 that will challenge both the American people and the world community.

Despite the complexities ahead, I remain an optimist. I am confident that if we collectively commit ourselves to a clean, healthy environment we can surpass the achievements of the 1980s and meet the serious challenges that face us in the coming decades. I hope that today's students will recognize their significant role in and responsibility for bringing about change and will rise to the occasion to improve the quality of our global environment.

The Visitors Center at the Mount Rushmore National Monument uses roof-mounted solar collectors to provide energy for heating and air-conditioning.

chapter 1

THE ULTIMATE ENERGY SOURCE

Throughout history humankind has looked up to the sun as a source of warmth and security. It is predictable, here now and forever, unlike other forms of energy, which are fleeting. Heat, light, chemical, radiant, mechanical, and nuclear energy—all are forms of energy, meaning they have the ability to do work. Energy cannot be destroyed, although it may be transformed into a usable form and dissipated. The energy used to operate any machine, therefore, follows a set pattern: It is converted into a more usable or convenient form and finally dissipated as heat. The rate at which energy is converted from one form to another is often as important as the total energy converted. This rate is referred to as power, or more simply, as the time it takes to do work. One way to measure power is the watt—the rate of work represented by an electrical current of one ampere under the pressure of one volt for one second. It takes 40 watts of power to light an ordinary 40-watt light bulb for 1 second. Because the watt is too small a unit of measure for most industrial uses, a unit called the kilowatt, equal to 1,000 watts, is used.

As the sun rises, the thin skin of life between the earth's crust and its enclosing atmosphere responds. Flowers open, reptiles gain the warmth that enables them to move, green leaves renew their daily task of turning sunlight into food. The variety of life-forms depending more or less on the sun's energy is enormous. These living things are small, varied, and complex, whereas the star they depend upon is relatively huge and simple.

THE SUN

The ball of fiery gases known as the sun has a mass many times greater than that of the earth. If the earth weighed 1 ounce, then the sun would weigh 10 tons. Although the sun seems relatively close to the earth by comparison to the outer planets, it is more than 93 million miles away. Sunlight takes 8 minutes and 20 seconds to reach earth; light from Proxima Centuri, the star closest to the solar system, takes more than 4 years to reach earth. Scientists study the spectrum of light, or the band of colors arranged by wavelength from shortest to longest, from the sun to learn about its elements. It is estimated that the sun is made up of 70% hydrogen and 28% helium. The other 2% of the sun's mass consists of traces of many other elements—carbon, nitrogen, oxygen, magnesium, and silicon.

The sun is an energy-conversion system in which matter becomes energy, as Einstein's equation $E = mc^2$ explained. By this process, a little mass creates a lot of energy. The system appears to be long-lived and self-regulating, functioning for billions of years within rather close limits. The outer edge of the sun also gives off heat and light. It has no surface as such; instead, it is made up of several stacked-up gaseous layers similar to the clouds in the earth's atmosphere. As is also the case with the earth, these outer layers are very thin compared with the diameter of the underlying globe. The core, or hot interior region, of the sun is where most of the heat generation takes place. Under the extreme force of gravity, the density of hot gas increases toward the center. And as with any gas in the earth's atmosphere, as the pressure increases, so does the temperature. Although no one has ever measured it, mathematical models suggest that the sun's energy-generating core reaches 27 million degrees Fahrenheit (F), with a gravitational force 250 billion times that of earth. The sun's innermost region produces heat and light by means of thermonuclear reactions. These complicated reactions occur when extreme heat and pressure in the sun's core cause hydrogen to be transformed into helium in a process called fusion. When nuclei combine, energy is released.

Part of the sun's matter is thus transformed into energy and released in the form of heat and light. The reactions in the interior are basically the same as the one that occurs during the explosion of a hydrogen bomb. The energy given off by the sun is equal to millions of hydrogen bomb explosions every minute. The earth is protected from this immense heat and radiation by its distance from the sun and by its sheltering atmosphere. However, that protection is not absolute, as humans' tendency to get sunburned shows.

The sun radiates energy in every direction. This radiant energy moves outward through space in electromagnetic waves that have a wide range of frequencies and lengths. One band of waves registers in the retinas of the eye as visible light. But there are many wavelengths that are both longer and shorter than the light we can see. The sun also shoots out vast streams of electrons

Parabolic mirrors collect solar energy to heat and cool the Information Center at the Palo Verde Nuclear Generating Station in Arizona.

that often interfere with communications systems on earth. These electrons create the phenomena known as auroras in the regions nearest to the earth's poles. The distance between the earth and the sun allows the earth to intercept just enough solar radiation to have nourished the development of life-forms. The earth's atmosphere screens out the dangerous short-wavelength radiation. Once radiant energy has been absorbed by molecules on the earth, it may be transformed into other types of energy. Nearly all the energy found in any form on the surface of the earth has been derived from solar radiation.

There are three mechanisms by which heat may be transferred from one place to another: radiation, conduction, and convection. Radiation involves the transfer of energy by means of electromagnetic waves. Conduction is the transfer of heat through a material by molecular motion. For example, if a rod is heated at one end, the molecules there vibrate faster, and that energy is transferred down the rod because of successive molecular collisions. Metals are good conductors of heat, but materials such as glass, ceramics, wood, and gases are poor conductors. Convection is the process by which a warm liquid or gas moves from one place to another, resulting in a transfer of energy. Radiation is the dominant transport mechanism for the sun's energy. In free space, radiation would travel a distance equal to the sun's radius in a little more than two seconds, but absorption is so strong inside the sun that it takes 1 million to 2 million years for the radiation to make its way out. Thus, the light and warmth the earth receives today were produced near the sun's center more than a million years ago. A radiation zone reaches out from the core for about 70% of the sun's radius. Core radiation slowly diffuses through this region until it reaches the boundary of the convection zone, where the temperature has fallen and made the gas highly opaque. Turbulent convection then takes over. In this process, the warm gases move from the interior to the outer layers, causing a transfer of energy, eventually carrying the flow of energy all the way to the sun's visible surface.

The sun's outer layer, or *photosphere* (from the Greek meaning "sphere of light"), is the one from which light emerges. Here the density of the hot gas is so tenuous that on earth it would be considered a vacuum. The surface itself is a thin shell only a few hundred miles thick. This shell, or *chromosphere* (meaning "sphere of color"), is made up of fiery red clouds of gases, which sometimes shoot as far as a million miles out into space. Around these two spheres is a more diffuse region called the corona, another area of gases that extends more than 9 million miles into space. Both of these layers are visible only during

This giant collecting dish at the Crosbyton Solar Power Project in Texas is part of one of the first systems to commercially market electricity produced by solar energy.

a total solar eclipse. The availability of these images has increased since the invention of the coronagraph, a device that can produce an artificial eclipse when fitted to a telescope.

The total amount of energy given off by the sun is immense. It is estimated that the sun radiates the heat equivalent of 400 billion trillion tons of burning coal every hour. The upper atmosphere of the earth receives 1 two-billionth of the sun's total energy, an amount equivalent to the energy represented by 126 billion horsepower. However, not all of the sun's energy that reaches earth's upper atmosphere penetrates to the surface. Thirty percent of the solar energy that reaches the earth bounces back directly into space, 23% is absorbed by the earth's atmosphere, and 47% is absorbed by the earth and converted into heat. Only 0.03% of the solar energy reaching the earth's surface is captured by plants and used in the process of photosynthesis, providing all the food energy for living creatures on earth and all the stored fossil-fuel energy.

The energy that sunlight provides through photosynthesis for plant growth is responsible for all the earth's food, both vegetable and, indirectly, animal. It also fuels the growth of wood, the burning of which supplies the energy needs of much of the developing world. Fossil fuels such as coal, petroleum, and natural gas are the partially decayed remains of earlier plant life, whose energy was once created by photosynthesis. The wind energy that drives windmills and sailing ships was originally created by sunlight heating the ground and lower atmosphere. Likewise, hydroelectric power is produced by water raised to high elevations by sunlight-induced evaporation and wind currents. Indeed, with the exception of the small fraction of energy generated in nuclear power plants (about 4% in the United States), all the world's energy is directly or indirectly supplied by the sun. The problem, as today's society has suddenly and painfully realized, is that the vast bulk of energy being used today is tapped from a very limited reserve of fossil fuels. These reserves were formed over a period of hundreds of millions of years and can be replenished only over a similar period of time. For all practical purposes, they are nonrenewable energy sources.

There are many attractive features of solar energy. First and foremost, it is available anywhere on earth and can be collected easily with portable devices. This advantage minimizes the problems of transporting fuel from where it is produced to where it is used to generate energy and of transporting power from where it is generated to where it is ultimately used.

The pumps, lights, refrigeration, and battery-charging equipment at this gas station in Bahrain are powered by roof-mounted solar panels.

Second, solar energy is nonpolluting. It is worth comparing this aspect with the adverse impact of burning fossil fuels. Besides contributing to acid rain, which can kill plant and animal life in aquatic environments, combustion of fossil fuels injects carbon dioxide into the atmosphere. This by-product contributes to the greenhouse effect, a gradual warming of the earth's atmosphere that many scientists say will eventually affect agricultural production and cause the oceans to rise, inundating coastal cities. A further advantage of solar energy is that "mining" it is environmentally safe. This factor is especially true when compared to coal mining, which is dangerous, polluting, and can be ecologically devastating, as in the case of strip-mining. Solar energy has its drawbacks, however. Sunlight is a rather dilute energy source. On a cloudless day in full sunlight the amount of solar energy striking the earth is about one kilowatt per square meter. Although the amount of available solar energy (allowing for clouds and nighttime) falling on just 0.2% of the United States is enough to fulfill all of this country's needs, it cannot be harvested at 100% efficiency. To produce solar energy equal to the total U.S. consumption of energy, at least 0.5% of the country's landmass would have to be used for collection. This mass would amount to about 18,000 square miles or one-sixth the area of the state of Arizona. A second major drawback is the inconsistency of direct sunlight as an energy source. Not only night, but clouds and seasonal changes affect the rate at which direct sunlight can be collected. Erratic sunlight creates the need for methods of storing the energy collected until it is needed. Similar requirements are imposed on solar-derived wind energy, which is also intermittent.

Most of the enormous amount of energy the earth receives from the sun is not used directly by humans. Although humans already benefit from the sun's natural heating of the earth, the potential uses of solar energy are almost unlimited. When the alternatives—decreasing the world's energy consumption or utilizing available energy more efficiently—are considered, it appears that solar energy sources may well be able to fill a large part of the gap that will be left by the depletion of fossil fuels.

This tandem bicycle, powered by a combination of human pedaling and solar cells, was developed by Dr. Ugur Ortabasi of the Solar Research Centre at the University of Queensland in Australia.

Painted black to absorb more solar energy, these aluminum solar collectors provide heat and air-conditioning for this suburban home in Maryland.

chapter 2

EARLY USES OF
SOLAR ENERGY

The sun has been so important to human life and civilization that it was a central element of many of the world's early religions and belief systems. Knowledge of the changing seasons alerted early peoples to the reappearance of edible plants and the migratory movements of game animals. With the mastery of agriculture, people came to rely on the sun's energy to grow food and on its regularity to plan growing seasons, predict the return of floodwaters, and, in general, organize their time and work routines to the best advantage. So vital was a knowledge of the sun's behavior that the early magicians and priests who worshiped it became the first astronomers. Religion and astronomical knowledge were closely tied together, and the first kings and religious leaders often were chosen for their supposed ability to influence the sun to produce good weather and good harvests. When they failed, they were often killed and replaced. In some early civilizations, the sun itself became a god. It is no wonder that ancient peoples shuddered in fear during a solar eclipse, for without a scientific understanding of what was happening, they must have thought that the sun had vanished forever.

One of the most fascinating ancient monuments to the sun is Stonehenge in southwest England. It is believed to have been built after centuries of study of the sun, and it appears to have been rebuilt many times over the span of seven centuries. Standing proudly after 4,000 years, it is the gathering place each year of hundreds of people who come to see the midsummer dawn. Stonehenge was designed so that the sun would rise directly over the Heel Stone, a small stone about 250 feet away from the 3 main arches. Because of the gradual shift in the earth's axis over the centuries, it now rises just to the left of the Heel Stone.

The remains of ancient observatories at the Maya ruins of Chichén Itzá in the Yucatán Peninsula and at Monte Alban in western Mexico suggest that these early civilizations attempted to determine the precise days when the sun's shadows vanished during the zenith passage—the period when the sun shines directly overhead and casts no shadows at noon—a phenomenon that occurs only in the tropics. In this culture, the zenith passage marked the onset of the rainy season and the beginning of planting.

The sun was vitally important to the ancient Egyptians, whose success in feeding themselves depended not only on the sun's energy but on the regular seasonal flooding of the Nile. Accurate calendars, developed and kept by the priests, were considered secret religious knowledge. Sun symbols and sun myths became an integral part of Egyptian culture. The daily course of the sun through the sky was interpreted as a journey made by Ra, the sun god, in his celestial boat, moving from east to west. At dusk, Ra exchanged the boat of the day for another to take him through Tuat, the underworld. Ra's voyage through the underworld was a perilous one, in which he did battle with the giant serpent of darkness, Apep. If he emerged victorious, Ra would again exchange his boat for the daytime vessel and sail through the sky.

THE GREEKS AND ROMANS

The Greeks were the first to make extensive use of solar architectural design. Greek houses were constructed of adobe walls one and a half feet thick on the northern side to keep out the chill of winter winds. The main living area, on the southern side, faced a portico supported by wooden pillars that led to an open courtyard. This design permitted sunlight to enter the home through the south-facing portico, and the radiation was trapped in the earthen floor and thick adobe. Both the earth and adobe provided a useful form of heat storage and allowed for a gentle release of solar warmth throughout the winter night.

The Greeks were also the first to use large curved mirrors that could concentrate the sun's rays onto an object with enough intensity to make it burst into flames. Made of polished silver, copper, or brass, the earliest mirrors consisted of flat pieces of metal joined together to form a roughly curving surface. These were later replaced with concave spherical mirrors made of single sheets of metal. The Greeks soon discovered that a mirror with a parabolic surface was even more powerful than a spherical one. Dositheius, a mathematician of the 3rd century B.C. who built the first parabolic mirror, discovered that solar rays bouncing off such a mirror are focused almost to a point. Because solar energy is concentrated into a smaller area with a parabolic mirror than with a spherical mirror of a similar size, a parabolic one produces higher temperatures. Legend says that in 212 B.C., Archimedes used such mirrors to set fire to the ships of the invading Romans at Syracuse.

The Romans took the Greek discoveries and improved upon them. The Roman architect Vitruvius, in his 10-volume work *De Architectura*, listed specific guidelines for orienting houses, public buildings, temples, and even whole cities to the sun. He stressed that the siting and construction of an individual dwelling, as well as the plan for a whole community, should fit both topography and climate.

Window glass was developed in Rome in the 1st century A.D. This invention made possible better and more effective solar design. In his writings on solar design, Pliny the Younger described the heliocaminus, or solar furnace. A room in his villa, with southwest windows glazed with mica, a crude form of glass, became extremely hot as the glass trapped the solar heat on the inside. When the same idea was applied to agriculture, the first greenhouses were built.

The Romans also applied solar design principles to their public baths. The hottest bath, the caldarium, faced south and had enormous windows along its southern wall. Solar heat was trapped and stored in the floors and walls during the day and released in the evening. The first laws protecting the accessibility of buildings to sunlight were also developed by the Romans. To this day, certain rights to unobstructed sunshine are recognized in Great Britain and other countries whose legal systems are rooted in Roman law.

After the collapse of the Roman Empire, Arab scholars helped to preserve the discoveries of solar design and concentrating mirrors while European science languished during the Dark Ages. In the 16th century, among his many other designs and fanciful projects, Leonardo da Vinci proposed a large mirror that would generate power for a dyeing factory. He suggested that the mirror be four miles across to heat a pool of water to boiling. In 1561, alchemists made perfume by submerging certain types of flowers in a water-filled vase placed at the focal point of a spherical mirror. The concentrated solar heat caused the essence of the flowers to diffuse into the water.

This 18th-century print shows an early French solar furnace. The French continue to pioneer in solar energy applications.

Early Uses of Solar Energy / 27

EARLY EXPERIMENTS WITH SOLAR ENERGY

In the late 18th century, the chemist Antoine Lavoisier experimented with solar furnaces in a search for a powerful source of heat. With a special 52-inch-diameter lens, filled with alcohol to increase its refractive powers, he was able to melt platinum at 3,236°F. The first multiple-mirror solar furnace was constructed in 1797 by the French scientist George Buffon, who used an array of 168 flat, square mirrors, each 6 inches on a side, to ignite a woodpile at a distance of 197 feet. The French have maintained an interest in solar furnaces ever since those early experiments and now operate a large research facility in Odeillo, France, in the Pyrenees Mountains.

Pioneering work with flat-plate collectors, which absorb solar radiation without the concentration of energy that occurs with curved mirrors, was done during the middle of the 18th century by Nicolas de Saussure, a Swiss scientist. He designed a solar oven consisting of glass plates spaced above a blackened surface enclosed by an insulated box. The sunlight entered the box through the glass and was absorbed by the black surface. By chemically coating the outside surfaces of the glass, de Saussure was able to achieve temperatures as high as 302°F. This simple approach is still the most inexpensive and reliable way to collect energy from the sun.

A type of energy-concentrating solar cooker was described in 1878 in an article by W. Adams of Bombay, India, in *Scientific American.* This apparatus was an eight-sided conical box lined with silvered glass mirrors that focused light through a cylindrical bell jar onto the food container. This invention worked extremely well. Later, Dr. Charles Abbot, a leader of American

A 19th-century French engraving showing a printing press operated by solar energy.

solar scientists best known for his early work analyzing the solar spectrum, developed a solar cooker that not only employed parabolic mirrors but also used a type of heat storage system that enabled food to be cooked after sundown.

Solar engines—devices that converted the sun's energy into mechanical energy—were developed in the latter part of the 19th century by Augustin Mouchot in France and John Ericsson in the United States. Both made a notable advance in collector design by introducing a circular cone reflector, similar in shape to a satellite dish antenna, that focused light uniformly along its axis rather than at a small spot, as had been done previously. In 1882, Mouchot, in collaboration with Abel Pifre, went on to develop a solar steam engine that operated a printing press.

John Ericsson, best known for designing the Union navy's ironclad ship, the *Monitor*, devoted more than 20 years to solar-powered engines. He built 9 such engines between 1860

Dr. Charles Abbot (right) displays one of the solar thermal energy systems he designed in the 1930s. The mirrors focused sunlight on tubes of volatile liquid, which produced steam to power a one-half-horsepower engine.

and 1883, but eventually concluded that solar engines were 10 times more expensive than conventional engines and, therefore, were not economically feasible.

SOLAR WATER HEATING

The heating of water for bathing was a regular practice in ancient Rome, and the power of the sun was often used. Typical Roman baths were fed from man-made channels left open to the sun and lined with grooved black slate so that the water was heated as it ran through the grooves. The practice of frequent bathing died out in the Middle Ages and did not return until the 18th century, when developments in personal hygiene increased the need for hot water.

The first solar water heater was a simple metal tank painted black and tilted toward the sun. In 1891, Clarence M. Kemp of Baltimore, Maryland, patented a way of combining metal heating tanks with enclosing boxes that allowed the tanks to collect and retain solar heat. He called his invention the Climax. This invention became the first commercial solar water heater, with models ranging in size from 32 gallons to 700 gallons. Each model contained four long cylindrical tanks made of heavy galvanized iron and painted a dull black. They lay horizontally next to each other in a pine box lined with felt paper and covered with a sheet of glass. The box was placed on a sloping roof or on brackets at an angle to a wall. The tanks were filled with water, which was heated by the sun, and they could be emptied before freezing so that the tanks would not split. From 1900 to 1911, more than a dozen inventors filed patents for improvements on the Climax. The best was one called the Day

Developed by Bell Laboratories, this panel of photovoltaic cells was installed on a rural telephone line in 1955 and made possible the first solar-powered telephone call.

and Night, invented by William J. Bailey in 1909. This heater's main selling point was that it could supply hot water during both the day and night. Bailey separated the heater into two parts, the solar heater and a water storage tank placed next to the stove in the kitchen to keep the water warm at night. Thin copper pipes and a metal absorber plate transmitted the solar heat accumulated in the hot box to the storage tank.

SOLAR ENERGY IN THE 20TH CENTURY

By the turn of the century, solar power was being used for a variety of purposes. A water pump, operated by the steam from a boiler heated by a large solar reflector, was built by Aubrey C. Eneas in 1901. He built the pump to provide water for irrigation on a Pasadena, California, ostrich farm. This pump produced 4 horsepower and cost only $2,500. On a typical day, the machine began operating about an hour after sunrise and continued to pump water until an hour after sunset. To start the motor, an attendant turned the reflector toward the rising sun. As the rays struck the mirror, the boiler heated up. The reflector automatically followed the sun throughout the day. Other water pump designs were also appearing during this era. In 1908, in Needles, California, H. E. Willsie and John Boyle, Jr., built a 15-kilowatt solar engine that ran a water pump, a compressor, and two circulating pumps. This device used a solar collector to heat water, which in turn was used to heat a volatile liquid, sulfur dioxide, which was then vaporized. The vapor drove a simple low-temperature engine. But the cost of construction was many times that of a conventional engine, and the project was considered a financial failure.

Frank Shuman was the first to come close to making a commercial success of a solar-powered engine. Using a flat-plate collector, he constructed a successful three-and-a-half-horsepower engine. Solar energy was absorbed by the chemical ether, which boiled to provide steam for the engine. Since the flat-plate collector did not have to track the sun, it was less costly and more simple than other designs. Shuman also constructed a solar power plant near Cairo, Egypt, which was the first to use

The houses in this village in Gansu province in China have been built and arranged to take maximum advantage of the sun. This kind of passive solar home heating was used by the ancient Greeks and Romans as well.

small motors to rotate reflectors to follow the sun. These motors were directed by heat sensors and thermostatic controls. But Shuman's system fell into disrepair during World War I.

One of the most practical applications of solar energy involves the production of fresh water. A solar still, built in the desert near Las Salinas, Chile, by J. Harding and C. Wilson, delivered 6,076 gallons of water per day and operated for about 40 years. Solar stills, manufactured for World War II survival kits to be used in the South Pacific, were operated by simply stretching a plastic sheet, held down by rocks and dirt, over a hole in the ground. Moisture in the ground rose and condensed on the plastic and dripped back into a container.

By the 1930s, interest in solar energy had spread to include home heating. M. M. Hottinger conducted solar experiments in 1935 at the Zurich Institute of Technology and found that solar radiation could be captured by collectors mounted on the south-facing roofs or walls of a building. By using water or air as the heat-transfer medium, he found he could store solar heat in insulated tanks of water or containers of rocks.

The International Solar Energy Society was formed in 1954 and in 1955 organized two international solar energy conferences. By the mid-1950s, the only successful application of solar energy was for space heating and water heating in Florida and California. Tens of thousands of solar water heaters were sold in both states by 1955, when cheap fossil fuels caused a decline in sales. At that time, there were more than 50,000 of these devices in Miami alone, but by 1970 they had all but disappeared. They are still used in Japan and Australia.

These double-walled glass tubes are placed at the focal points of parabolic solar mirrors. Water flowing through the inner tube is heated to produce steam. The layer of air between the glass walls is an excellent insulator and increases the efficiency of heat absorption.

chapter 3

COLLECTING SUNSHINE

The sun's energy can be used in two ways. First, it can be used to heat a fluid, such as water or oil. Once heated, the fluid is transferred to another location to perform work, such as to heat a house or to create steam to drive an electrical turbine. Second, the development of new semiconducting materials has made it possible to convert light energy directly into electrical current. The heating of fluids is the oldest and most common method of using sunlight and requires that the sun's energy be collected and concentrated.

The most common means of collecting the sun's rays is with a flat-plate solar collector. There are many different designs, but the basic structure consists of a flat metal plate, painted black to absorb heat, built into an insulated box or container covered with glass. Pipes run into the box and are attached to the black plate so that heat travels by conduction through the metal and into the heat-transfer fluid. The glass cover allows visible light energy from the sun to enter the box but traps the infrared heat emissions from the metal plate, like the glass panels of a greenhouse, raising the efficiency of the collector. The heated water in the pipes is pumped either to a storage tank or to a traditional hot-water-heating system using radiators or baseboard units. Normally, several flat-plate collectors are connected in a series. In a conventional flat-plate collector, water is by far the best heat-transfer liquid because of its high heat capacity. Water can absorb a great deal of heat energy before its temperature is raised. But oil, solutions of antifreezelike liquids, and even ordinary air are sometimes used to transfer heat, depending on the application.

A flat-plate collector's efficiency depends on a number of factors, first and foremost the amount of *insolation,* or solar energy, hitting it, which in turn is affected by the collector's orientation and tilt. A collector facing directly into the sun will receive the most insolation. Whereas pointing the collector toward true south is the best orientation in the Northern Hemisphere, just west of south is also good. For winter heating, the tilt of the collector must be increased, because the sun is lower in the sky in the winter. The generally accepted procedure is to add 15° of tilt to the local latitude to get the optimum total angle from the horizontal position for best winter heating; for the summer, 15° should be subtracted from the local latitude. For year-round use, the angle of tilt should equal the angle of local latitude.

Also important is the design of the collector and the rate at which it loses heat to its surroundings. Heat loss is directly proportional to the difference between the ambient, or outdoor, temperature and that inside the collector. As one would imagine, more heat escapes when the air outside the collector is very cold. The large surface area of the flat collector plate itself also encourages heat loss. A vacuum may be used as an insulator to reduce this kind of heat loss. In these more advanced evacuated tube designs, a double-walled tube holds a vacuum between two layers of glass. A copper pipe replaces the flat-plate collector and carries the heat transfer fluid through the inner glass tube. Each unit has a U-shaped design. A variation on this design uses a refrigerant such as freon that continually vaporizes as it is heated. The gas rises to the top of the tube, where it transfers its heat to circulating water before condensing and falling back to the lower end of the tube.

Two major drawbacks to evacuated-tube collectors are the tendency of the tubes to break and the high cost of replacing them. Although harder glass and better manufacturing processes have reduced breakage, this system is still more expensive than the regular flat-plate collector.

By the early 1980s, these solar collectors were producing only about 0.01% of the world's energy supply. But their use is increasing. Japan, which currently has more than 3 million of them on the roofs of its houses, plans to equip 20% of its homes with solar collectors by the mid-1990s, hoping to save 5 million metric tons of oil a year. Currently, more than half the houses in Israel are equipped with solar collectors, and Canada and Australia are aiming to use collectors to generate 12% of their energy by the year 2000.

LINEAR FOCUSING SOLAR COLLECTORS

Another type of solar collector is the linear focusing collector, commonly known as the parabolic trough. A typical model has a parabolic mirror suspended beneath a black pipe through which some sort of heat-transfer liquid flows. The troughs are oriented from the east to the west, so the collector need not track the sun; however, the angle does need to be adjusted every month or so.

In this type of collector, sunlight is focused on the pipe, which is filled with oil or pressurized water running the length of the collector. A variation of this system uses a small sensor that notes the position of the sun and relays the information to a series of electronic motors that keep the collectors correctly aligned. But they can only track the sun on a single axis, east to west. A more sophisticated design for a parabolic collector uses the system of point focusing whereby the mirrors, usually in the form of dish collectors, move along two axes by means of a computerized mechanical guidance system and concentrate the sun's direct light onto one central absorber. This type is more efficient than are collectors using immobile mirrors. The first large parabolic collectors were built in the late 1960s by the French at Odeillo. They used 63 large computer-controlled heliostats, or sun-tracking mirrors, spread over 10 acres. All sunlight was concentrated by 138-foot-wide parabolic reflectors onto a receiver, which was able to reach temperatures of more than 5,400°F. This intense heat was used to melt certain metals for industry. Heat from smaller, less intense systems is now used in oil refineries, breweries, and even potato processing plants, especially in the United States. A recent installation in Nio, Japan, uses a bank of 2,500 flat-plate collectors to track the sun and more than 120 parabolic trough reflectors to collect radiation and further concentrate it.

SOLAR THERMAL POWER

Solar thermal power plants using steam turbines currently seem most likely to be the first large-scale solar electricity producers. The technical feasibility of solar thermal systems was established at the turn of the century. The solar thermal power systems that are under development today are based on the same principles that were used in those early applications. Any improvements are mainly the result of modern design and manufacturing techniques. The development of efficient thermal devices to produce electricity usually requires some sort of optical device, which calls for a large initial equipment expenditure. This cost is the dominant factor in determining the cost per kilowatt hour of solar-derived electricity, since solar fuel is free. The only real pollution problem attached to solar thermal power is the disposal of waste heat, and solar energy is no different from other sources of energy in this respect. Solar power plants must also be located at a considerable distance from major population centers, so transmission losses and costs must also be considered. Perhaps the most troublesome feature of solar electric plants is the fluctuation in available solar energy resulting from clouds and nighttime. Therefore, an adequate means of energy storage is needed. Solar thermal power can be generated using either the central receiver system or the distributed collector system.

SOLAR POWER TOWERS

A central receiver, also known as a solar power tower, uses a heat-exchange fluid, usually water, held in a receiver

Solar One near Barstow, California, went operational in 1982. More than 1,800 individual mirrors, or heliostats, focus sunlight on a steam boiler atop the 250-foot-high central tower, producing electricity for 1,500 homes.

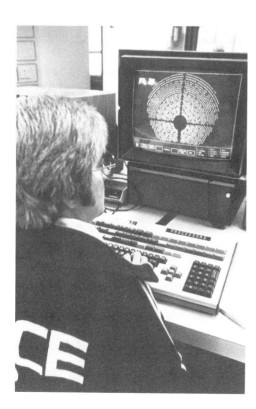

From his console a technician monitors the performance of Solar One's heliostats, which are controlled by computer and automatically track the sun as it crosses the sky.

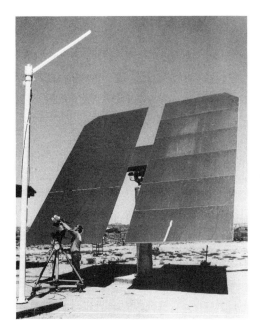

One of Solar One's individual heliostats undergoing testing before installation. The white post on the left carries a sensor that determines how well the automatic tracking system is working.

composed of a series of blackened pipes located in the top of a tall tower. The fluid is heated by solar energy reflected from thousands of individual heliostats. The focused solar radiation is absorbed by the receiver and converted to heat, which is used to vaporize the circulating water. The high-temperature, high-pressure steam is returned to the ground and used to power an electric turbine generator and, simultaneously, to deliver energy to a thermal storage unit for future use. A 100-megawatt solar electric facility would require about 24,000 heliostats spread out over an area of 1.8 square miles if located in a central California desert region. The tower for a plant of this size would have to be at least 858 feet high, and the temperature achieved at the focus of the mirrors on the receiver would be between 1,112° and 2,192°F; therefore, the central receiver approach is best suited to large-scale applications. The first large-scale central receiver system was built in Odeillo, France. Another was added at nearby Tavgasonne; both were 1 and 2 megawatt plants, respectively. In the 1970s, a number of pilot plants were built, including the 10-megawatt Solar One plant at Dagett, near Barstow, California, in the Mojave Desert, where there are 300 cloudless days a year. Solar One consists of 1,818 heliostats and cost $142 million to build, most of which was supplied by large aerospace and power companies. The largest current project is the European Phoebus project, a 30-megawatt facility under construction in Jordan. Because the heliostat field makes up almost half of the total cost of a solar tower system, good design is crucial. The heliostats must be spaced in such a way as to avoid shading and blocking of the reflected radiation. The responsibility for design fell to aerospace companies, whose experience in designing highly accurate tracking mechanisms for ballistic missiles was applicable to this work.

Critics are quick to point out that solar power towers can only be feasible when located in deserts. And, they ask, with the immense cost, might these plants be abandoned in favor of better facilities in the near future? Finally, they argue, the cost of the electricity produced is much too high and has to fall before the solar power tower can be economical. One solution to this dilemma is to place a solar power tower next to an existing fossil fuel power plant. These conventional plants could be "repowered" using solar thermal energy, where the solar heat could be used to power the normal steam generators. On days when too little solar power was available, oil, coal, or gas could be used. A survey taken by the Department of Energy in 16 southwestern states found that about 13,000 megawatts of solar power could be generated in repowering schemes. This rerouting would save 11% of the oil and gas currently consumed by power companies. Solar power towers could also be used in the same way in industry to provide heat. Four existing power stations, the largest in El Paso, Texas, have been repowered using solar power towers. It is estimated that the towers have been able to produce up to half of the station's steam requirements.

DISTRIBUTED COLLECTORS

Distributed collectors use a moderate degree of concentration and depend on collecting solar energy by using the sun to heat a fluid to a moderately high temperature, usually from 212° to 932°F. Distributed collectors operate by collecting sunlight and converting it to heat at each individual collector module. Each module has its own heat collection pipe equivalent to the central receiver in the solar tower. Collection areas are typically a few square meters, so that many thousands of these modules are required to provide the power output equivalent to a 100-megawatt solar tower. Because there is a large amount of interconnected plumbing required, the distributed collector approach can only be used for smaller installations. This concept is becoming increasingly popular for small electric power plants and for systems that provide both heat and power.

A system of distributed collectors lends itself easily to a wide variety of low-power on-site applications. This system can be used to pump water for irrigation, to produce steam for industrial purposes, to generate electricity in small- and medium-sized installations, and to supply heat for residential use. The best application of this concept is in remote areas whose geographic location makes obtaining conventional electricity both difficult and costly.

Between 1984 and 1988 the LUZ Corporation, based in Los Angeles, installed several commercial solar thermal electric plants having a total generating capacity of 275 megawatts in the Mojave Desert. This system uses mirrors mounted in parabolic troughs to focus sunlight on oil-carrying receiver pipes. The oil is heated as it circulates through the pipe and is used to create steam that drives a turbine generator. Another such facility, with a capacity of over 300 megawatts, is under construction at Harper Lake in southern California. This facility is actually a series of

Near Daggett, California, a 66-acre expanse of 560 solar parabolic mirrors produces 13.8 megawatts of commercial electricity.

At the Comprehensive Indian Health Care Facility at Ada, Colorado, more than 1,100 solar collectors provide heat, hot water, and air-conditioning for the 75-bed hospital.

80-megawatt plants, each consisting of 852 100-meter-long, independently operated solar collectors. Construction time for the first of these plants was 9 months, far less than the 6 to 12 years typical for a conventional power station.

Although solar collectors of all types, including solar power towers, have been successful in the past, hopes for the future of solar energy are being placed on photovoltaic power cells.

This new solar cell developed by TRW engineers in Redondo Beach, California, combines the concentrating power of a curved reflector with the direct electricity generation of a gallium arsenide photovoltaic cell.

chapter 4

PHOTOVOLTAIC CELLS

The utilization of solar energy for water heating and space heating is now well established, but the conversion of solar energy into electricity is not as well developed. The various systems for using solar reflectors to heat water, oil, or other volatile liquids to create steam to drive electric generators have been shown to be technically feasible, but the cost of this type of electricity is still greater than the cost of electric power produced by fossil fuel power plants. Solar thermal power plants are now used experimentally or to supplement other sources of electric power, but they are not yet ready to take over basic electricity production anywhere in the world. However, another solar energy technology is beginning to show this kind of potential—the direct generation of electricity through *photovoltaic cells*, or layered materials whose electrons are activated by light.

Photovoltaic cells, if they can be made economically efficient, would be the perfect energy source, creating electricity with no pollution, no noise, and often with no moving parts. Photovoltaic systems need minimal maintenance and do not use water for either cooling or steam generation; they are therefore well suited to remote or arid regions. They can also operate on any scale, from portable models of just a few watts for remote communications to full-scale power plants covering millions of square meters and generating many megawatts.

THE PHOTOVOLTAIC PROCESS

Light energy from the sun arrives in the form of photons—small packets, or particles, of energy—a fact discovered in the early 20th century by the German physicist Max Planck. Planck named the packets of energy *quanta,* and he won a Nobel Prize for his efforts. It was known that when these photons or quanta of light hit certain substances, they could knock electrons out of those substances. Albert Einstein won a Nobel Prize for discovering how the wavelengths of light determined the "punch" of these energy packets—their ability to knock electrons out of the atoms of certain substances.

The solar, or photovoltaic, cell takes advantage of this process by exposing masses of electrons to solar photons. It consists of sandwichlike layers of materials called semiconductors. These substances transmit electric current, but not as efficiently as metallic conductors. Common semiconductor materials include silicon, germanium, cadmium, and gallium, with slight additions of "impurities," such as arsenic, boron, phosphorous, or sulfur. One layer of the cell may have a germanium or silicon base with an extra layer of phosphorous or arsenic painted on it, giving it an excess of loosely bound electrons. It is known as the n-type layer. The other layer is a base material coated with a thin film of boron, which contains an excess of positive charges or "holes" toward which excess electrons might be attracted. This is known as the p-type layer.

A typical photovoltaic unit, or module, consists of 20 to 40 of these two-layered cells. As photons of light energy strike the n-type layer that faces the sun, millions of electrons are knocked out of their atomic orbits and are attracted to the positive charge of the p-type layer, causing electric current to flow. As current flows through a series of such cells, the voltage continues to build.

Modules of photovoltaic cells are connected to form panels that are assembled into groups called *arrays*. The actual current and voltage produced by the array depends on how the modules and panels are wired together. The array is connected to a battery that both stores electricity and sends it out at an even

High in the Sierra Estrella, a mountain range near Phoenix, Arizona, this array of photovoltaic cells generates power for a radio communications network that makes possible the worldwide use of cellular phones.

rate. The battery is a vital part of the process, as electric current from the array can fluctuate from moment to moment as clouds pass overhead.

Researchers continue to find new materials that will convert sunlight into electricity even more efficiently. The present generation of photovoltaic cells only converts one-sixth of the insolation, or solar energy, that falls upon the cells. With current technology, it would take an area of 6 square miles covered with photovoltaic cells to replace a 2,000 megawatt power station.

More economical solar cells are now being produced using techniques developed for the printing industry. Using these techniques, a thin film of silicon is laid down on sheets of stainless steel, which roll off an assembly line. The cells made from these sheets cost only one-tenth as much as earlier generations of solar cells cost. These mass production processes could make photovoltaic cells competitive with other means of electric power production. The Japanese have found a way to print photovoltaic cells directly onto a glass base using cadmium sulphide, cutting the cost of these cells by more than half. Another material used to manufacture solar cells is iridium phosphide, the same material used to make light-emitting diodes (LEDs). The electronics industry pioneered LED technology, which converts electricity into light, and is now trying to perfect the reverse process.

New solar cells being developed in the laboratory promise to convert sunlight into electricity with an efficiency rate of 30% or more. This conversion is possible when two different types of solar cells are sandwiched together. A translucent gallium arsenide cell is placed on top of a silicon cell. The gallium arsenide cell absorbs the visible light; the silicon cell below is activated by infrared light that passes through the top cell. The idea behind stacked cells is that two or more photovoltaic materials can absorb light more efficiently than can a single material. The cells are mounted beneath arrays of lenses that concentrate the sunlight. This system works at very high rates of efficiency, but at present it is too costly for commercial use.

Mounting inexpensive solar cells under cheap acrylic concentrating lenses has also been tried. This method increases the amount of electricity that the cells produce, but the lenses require motorized sun-tracking systems that add to the cost, making this technique a viable solution only for large-scale applications in areas such as the southwestern United States.

Pilot photovoltaic power plants built years ago are showing that solar cells can work for extended periods. One such plant is Arco Solar's 6.5-megawatt Carinna Plain solar power station near San Luis Obispo, California. The world's largest photovoltaic plant, it produces more than 13 million kilowatt-hours of electricity a year, enough for more than 2,300 homes.

THE EVOLUTION OF PHOTOVOLTAIC CELLS

The principle of photovoltaic power—that certain substances produce electricity when exposed to light—has been known since the end of the 19th century. The first solar cells were made in 1889 by an American inventor named Charles Fritts. These cells were the size of a quarter, made of selenium and covered with a translucent gold film.

The first practical solar cells were made in 1954 by scientists at Bell Laboratories. They found that crystals of silicon could turn sunlight into electricity. The real push to develop photovoltaic cells came soon afterward with the growth of the U.S. space program, when a lightweight, independent power source was sought for satellites. Because the sun shines 24 hours a day in space, solar cells proved to be the perfect solution. Photovoltaic cells were used first on the *Vanguard* satellite launched in 1958 and have been used on thousands of satellites and spacecraft since. However, their high cost has ruled out their use for all but these special applications.

The rise of oil prices in the 1970s forced the U.S. government to develop policies that encouraged solar energy research, although these policies were reversed again in the 1980s. In 1982, the cost of producing electricity from solar cells was $1 per peak watt (peak refers to the amount of power produced on a clear day at noon, when the sun's rays are strongest). That cost has now fallen to only 20 to 30 cents per peak watt, but this rate is still about 5 times the cost of electricity from conventional energy sources.

In 1982, a photovoltaic power plant was built at Natural Bridges National Monument in southeastern Utah. Power was needed for the visitors' center and the auxiliary buildings at the site, and because the nearest utility power lines were inaccessible, the only options were a diesel generator or solar photovoltaics. The latter was selected, and an array, consisting of 250,000 cells, was constructed to supply electricity to all structures, including two 750-foot-deep water pumps. The array covers 1.5 acres and wraps around desert pine trees and outcroppings of rock. The system is capable of producing 100 kilowatts of electricity, although the complex only uses about 10 to 15 kilowatts. Excess power generated on sunny days is fed into 28 batteries, which can store electricity for at least a week of inclement weather.

The European Community (EC) also has been developing a program for 15 photovoltaic power plants throughout the Continent and the British Isles. So far these plants power traffic control equipment at the Nice Airport in France, drive water pumps in Belgium, and provide the entire electricity needs of the inhabitants of a small island off southern Italy. The governments of Italy and Spain are funding smaller systems for homes in remote areas, and Italy is also planning larger systems capable of generating up to 3 megawatts.

The largest photovoltaic plant in Europe is at Pellworm, a tiny island in the North Sea, 10 miles off the coast of Germany. It produces 300 kilowatts for about 100 people and as many sheep and provides power for the island's health spa. The plant has more than 17,500 solar cells spread across an area equal to two football fields. The solar panels are mounted above the ground so that the sheep can graze under them. Excess energy is stored in a network of batteries for use on nights or cloudy days. It is estimated that photovoltaics will provide up to 10% of Europe's electricity needs by the year 2025, according to a study conducted by the European Community.

A major study conducted by the Electric Power Research Institute in the United States suggested that solar cells offer considerable promise of becoming a source of bulk electric power around the world by the turn of the century. It is estimated that more than 50 million of the world's poorest families in the Third World live on less than 125 million acres of land, or less than 1

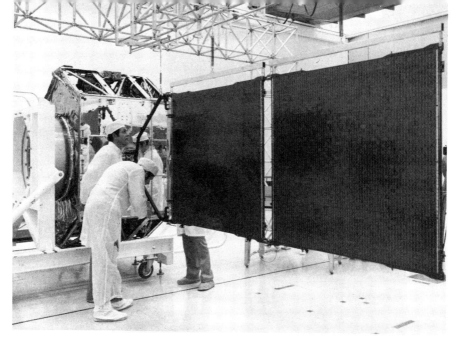

Photovoltaic cells are an ideal source of power for orbital satellites and spacecraft.

hectare per family, in fertile alluvial valleys, deltas, and on coastal plains. Many of these areas experience extended dry periods and need irrigation to cultivate crops. Large-scale irrigation is beyond the means of small farmers, but small solar powered water pumps might provide the answer.

Photovoltaic systems are already helping to fight disease. Vaccines decompose unless they are kept refrigerated, and conventional refrigeration units are not practical in some circumstances because they need a constant supply of electricity. But mobile refrigeration units, powered by an array of solar cells, are now available to transport vaccines to remote areas of the world.

Hybrid solar energy systems, in which a low-cost fossil fuel system is added to a photovoltaic system to compensate for variations in sunlight, have been suggested. A combined solar–natural gas hybrid would be one of the most environmentally benign ways of using fossil fuels to generate power. Hybrid mini-utility systems containing photovoltaic cells, batteries, and diesel generators already provide reliable power in some areas. A mini-utility system of this kind serves a small community on Coconut Island in the Torres Strait between Australia and New Guinea. Similar systems are being considered for Africa 1000, a project aiming to provide electricity to 1,000 African villages, and for the Australian outback.

RESIDENTIAL USES FOR PHOTOVOLTAIC CELLS

Units of photovoltaic cells of various designs can be installed in individual homes in the same way that solar thermal heating systems are installed. A typical residential installation might consist of an array of solar modules, a device to change the direct current produced by the cells to alternating current, and a group of batteries to store the electricity or a connecting unit to feed excess electricity into the town grid. Since a single cell generates about one-half kilowatt per square foot of cell surface, it would take an array of interconnected modules covering 550 to 860 square feet to generate enough electricity to meet the needs of a single-family home.

The design of a residential system must take into consideration geographic location, availability of local utility power, and the amount of electricity required. The modules must be mounted so that air can circulate underneath them to cool the cells. The dark anodized frames of today's photovoltaic cells blend well with most roof surfaces and do not detract from a home's appearance the way earlier systems did.

Workers in Harmarville, Pennsylvania, assemble a model solar house using glass panels to absorb solar energy.

chapter 5

THE SOLAR HOUSE

There are two types of solar energy systems used to heat and cool houses, *passive* and *active*. A passive solar energy system is one designed to use the entire building—walls, floors, windows, and roof—to collect, channel, and circulate heated or cooled air. It uses no mechanical or electrical devices, such as motors, pumps, or fans, to distribute heat. It depends instead on the natural processes of convection, conduction, and radiation. A passive solar structure is built and landscaped so that it becomes, in effect, one large solar collector. It has only one moving part— the earth turning under the sun.

Such a building requires an intelligent design and the right materials and components. To make a passive solar house even more energy efficient, it can be partially buried or constructed of thick adobe. Thick walls and floors made of stone, concrete, or brick serve to store heat. A top tier of windows can be used to channel heat to the side of the house away from the sun. Large, well-insulated, multipaned windows let sunlight in, and vents and open design allow warm air to circulate. In the summer, underground vents bring in cooler air, and trees may be used to block some of the sun's rays.

An active solar energy system employs large roof collectors that permit the sun to heat either air or a fluid that is pumped to a heat storage unit. The heat reservoirs are usually placed in basements or crawl spaces because of their large size and weight. An active system requires mechanical pumps and so cannot operate independently of other energy sources. Arrays of solar cells are sometimes used to provide additional heat or electricity. Active houses must face south so that the roof collectors can work efficiently.

HEAT STORAGE

One of the biggest problems with solar house design is heat storage. More heat than a house can use is available when the sun is shining. Some method must be employed to store the heat for use on cloudy days and at night. By storing excess heat, an active solar energy system can provide energy as needed at all times.

If cost were not a factor, a heat storage unit large enough to carry a house through the longest periods of sunless weather could be designed easily. For example, a 15,000- to 20,000-gallon storage tank in the basement could store enough heat to last through the winter. The cost of a heat storage unit is determined by the storage medium, the type and size of the storage container, the location of the container, and the need for heat exchangers, pumps, or fans to move the heat-transfer liquid.

Heat can also be stored in materials that change form, such as water, paraffin, or salt crystals. These materials can

This home in Stow, Massachussetts, is designed with south-facing glass windows to absorb the sun's energy. Built partially underground, the structure uses the surrounding earth to store heat in the winter and to cool the house during the summer.

absorb or release large amounts of heat as they change state from solid to liquid to gas and back again. Using a material with a large heat capacity can drastically reduce the total volume of the material required, but imperfect resolidification has limited its use.

Water is cheap and has a high heat capacity. A relatively small container of water will store large amounts of heat. Only one or two gallons of water are needed for each square foot of a flat-plate solar collector. There are problems with water, however. Storage tanks are costly and suffer from corrosion and leakage. In the past, only galvanized metal tanks were used, but now fiberglass and glass-lined tanks are used frequently despite their higher cost. Tanks located underground or in a basement are difficult and expensive to replace.

Rocks are the best known and most widely used heat storage medium for hot air systems. Many thousands of pounds of rocks, with diameters of one to four inches, are required to heat a home because of the low heat capacity of rock. Air, warmed by the sun, is pumped from collectors on the roof or in the yard to the bin of rocks, which absorbs the heat. To distribute the heat, air is either blown past the rocks or allowed to circulate through them by convection. The air then carries the heat to the rooms using fans or blowers. Six to eight inches of fiberglass insulation is needed around air ducts to prevent heat loss.

Some heat can also be stored in a building's walls and floors. If room temperatures are kept between 55° and 65°F over the course of a day, walls and floors can absorb excess heat on warmer days for evening or nighttime use. Depending on the building material, the expanse of south-facing glass, and the temperature difference between indoors and outdoors, fairly large amounts of heat can be stored. Concrete and stone are two of the best materials, since they store almost twice the heat as pine boards. A 6-inch concrete floor slab under a 1,200 square foot house can store almost enough heat to carry a well-insulated dwelling through a night of freezing temperatures. Brick and adobe can also be used. Whichever material is used, the floor should be exposed to as much direct sunlight as possible and should be well insulated around its edges.

South-facing masonry walls can be used for heat storage and as collectors, if a double layer of glass covers the outer surface, leaving an air space between the glass and wall. A series of ducts along the top of the wall permits heated air to rise between glass and wall into the room behind the wall. Another series of ducts along the bottom of the wall allows cooler replacement air to flow into the space.

SOLAR COOLING

Normally, the air in a building is cooled by elaborate electrically driven machinery, providing a comfortable environ-

This diagram shows the basic principle behind passive solar home heating—creating a circulating air flow to carry warm air throughout the house.

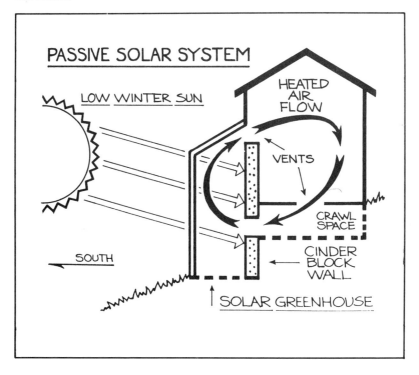

ment for the occupants. But producing electricity uses up fossil fuels and creates air pollution. The refrigerants in air conditioners can also damage the ozone layer. Solar cooling, however, avoids these problems.

Natural air conditioning can be accomplished with a passive solar heating system. Whereas a south-facing wall will store heat for the winter, cool air from the shaded north side of a building can be drawn into the living space and, after it is heated, withdrawn to the outside through the vents in the south wall.

If a solar heating system uses rocks to store heat, cooling can be accomplished with a device known as a rock bed generator. During the summer, when the rock bed is not being used to store solar heat, cool night air can be passed through an evaporative cooler and then used to cool down the rocks. During the day, ventilating air can be drawn through the rock bed, cooled, and circulated through the house.

THE TYPICAL SOLAR HOUSE

The typical contemporary solar house incorporates many special design features and architectural innovations that make the most efficient use of the heat energy available. Even little adaptations, such as overhangs that shade windows and walls against the high summer sun, are important. In winter, when the noonday sun is more than 40° below its highest summer position, the sun's rays can enter the structure. This simple architectural feature was fully exploited by the Greeks and Romans and is still an energy saver.

A solar house has a light-colored roof to reflect summer heat, double-paned glass to hold in heat during the winter, and windows on the south and east sides but few or none on the north and west sides. As sunlight passes through the glass, it strikes various surfaces, which absorb the light energy and radiate it back into the rooms in the form of heat, causing a greenhouse effect. Instead of the normal two-by-four studding, wall studs are two inches by six inches to provide space for more insulation.

Donald Finkell, an architect with the Tennessee Valley Authority, built this solar-heated doghouse for his dog Sampson.

This house in New York's Hudson River valley region features roof-mounted photovoltaic cells and was the first house in the United States to be completely solar powered, without connection to any utility power grid.

The most important building concept is the orientation of the house to the south to take advantage of the sun during the winter. Unfortunately, many builders ignore proven principles of energy conservation in favor of economic considerations. They will build houses facing the direction that allows them to get as many units on a tract of land as possible. It appears simpler and cheaper, in the short run, to build this way than to use innovative designs that are more energy efficient.

A solar house need not be complex or expensive to build. Most solar homes use a combination of active and passive systems, except in warmer areas, such as the U.S. Sunbelt, where passive systems are more than adequate.

COMMONLY ASKED QUESTIONS ABOUT SOLAR HOUSES

1. How is it possible to cool a house heated by solar energy?

Active solar energy systems, using either parabolic mirrors or photovoltaic cells, can be used to produce electricity, which will run any conventional air conditioner. With passively heated solar houses, which use south-facing glass windows and work like greenhouses, cooling is a matter of intelligent design. Cooler air from outside the house can be brought in through deep, cool, underground tunnels. One or more rooms may even be built partially underground. Roof vents can be opened in the summer to allow warm air to escape. If deciduous trees, those that seasonally shed their leaves, are planted in front of south-facing windows, in summer they will shade the house from the sun, but in winter the barren branches will not block sunlight. Awnings and shades can be used to advantage, and open interior windows and fans keep air moving. Pools of evaporating water on roofs will help in dry climates.

2. With all that glass exposure—and vents and windows—are solar houses prone to lose heat energy?

Yes. Insulation to prevent heat loss is very important in the design of a solar house. South-facing windows are generally triple-paned and heavily caulked. North-facing walls and attic floors are heavily insulated. Outside vents must be shut tight when not in use. Such houses are often called superinsulated houses. In addition, a dehumidifier may be necessary to prevent the buildup of excess moisture in such a tightly sealed environment, and an air purification system may be required for people sensitive to dust, pollen, or molds.

3. Which is better, a passive or an active solar home-heating system?

It depends. Passive systems have high initial construction costs, but low maintenance costs and high energy savings. Passive systems are ideal if built into the design of new houses. It is easier to add on an active system to an existing home, simply by installing photovoltaic cells on the roof, but these cells will have to be replaced at high cost some time during the life of the house, and energy isn't available at night or on cloudy days. Both systems will have problems in areas where housing is tightly clustered or where neighbors may build structures that block sunlight. Geography and climate determine how much sunlight is available in a particular locale, and how efficient such systems will be. Many installations will require small, supplementary, conventional heating systems. The relative cost of such systems will also depend on what happens to the price of oil and other fossil fuels in the decades to come.

Often passive and active solar systems will be used together. A passive design will provide home heating and cooling, and an active system, such as a roof-mounted flat plate collector, will provide water heating. The active system can also contribute to home heating if the heat it collects is transferred to insulated tanks containing water, rocks, or new high-tech materials such as magnesium hybride, all of which store heat well. Ther are already more than 1 million active solar water heating systems installed in American homes, saving an estimated 3 million barrels of oil. Such systems are also in wide use in Israel, Japan, and Australia.

At a 1989 World Energy Conference in Montreal, Canada, young environmentalist Viviane Saint-Aubert of France shows how to cook an egg using a parabolic solar reflector.

chapter 6

SOLAR ENERGY TODAY

Most solar energy applications can be divided into three categories: those used to heat water, those used to heat and cool homes and offices, and those used to generate electricity. Although water heating is still the number one use, both commercially and residentially, new applications are being developed all the time, and some of these will be of particular interest to people living in remote areas and to those in the Third World.

Scientists predict severe shortages of fresh water by the 21st century because of pollution. Some areas of the world have always had shortages of drinkable water. One way to help these areas is by the *solar distillation* of ocean salt water, in which the sun is used to produce fresh water for drinking, irrigation, and industrial use, with a by-product of salt. Solar distillation may in fact be the only answer to the water shortage in many remote parts of the world.

The technology is quite simple. Salt water is collected in shallow ditches or tanks covered with slanted pieces of glass or plastic. When heat from the sun is absorbed in the ditch or tank, the water becomes warm enough that it begins to evaporate. The water vapor rises until it reaches the glass or plastic cover, which is cooler than the water. The difference in temperature causes the vapor to condense, or to turn into liquid again, on the underside of the glass or plastic sheet. Distilled water runs down the slanted cover into collecting troughs. The water is fresh and pure because the salt, which evaporates at a higher temperature than the water, is left behind in the tank.

All such solar stills work on the same principle no matter how sophisticated they are. The world's first large-scale solar distillation plant was built in 1874 in Las Salinas, Chile, in the middle of the Atacama Desert. Its purpose was to supply water for potash miners and for the donkeys that transported potash from the mines there. The solar still was 51,000 square feet in area and produced 6,000 gallons of drinkable water every day until 1914, when fresh water was finally piped in.

Solar stills are functioning today in warm and sunny parts of the world. At Coober Pedy, Australia, drinkable water was hauled in by truck until 1966, when a still was constructed that produced 2,895 gallons of pure water daily from saline well water. The Leslie Salt Company of California uses the natural process of evaporation to produce about a million tons of salt a year on 29,000 acres of salt flats north of San Francisco.

SOLAR COOKERS

Another device that uses the sun's energy is the solar cooker. Numerous stoves and enclosed ovens have been designed that can concentrate enough heat to roast meat, boil water, and bake cakes. They combine the basic designs of the flat-plate and focusing collectors. Some operate by trapping heat

In Bonn, Germany, a policeman looks on with amusement as a member of the Green party arrives at Chancellor Helmut Kohl's office on his solar powered tricycle.

under glass, whereas others concentrate the sun's rays on the food. Though not yet mass-produced, they are being introduced in many tropical locales as a clean, cheap, portable source of heat for preparing food. The only drawback is that they cannot store heat for a rainy day.

THE SOLAR CHIMNEY

One of the most unusual experiments in solar energy has been going on in Manzanares, Spain, since 1982. A 656-foot-high, 33-foot-wide solar chimney was constructed to produce electricity. The core of the chimney is a central tower made of corrugated steel surrounded by a solar collector in the form of a transparent greenhouse. Air inside the greenhouse is heated by the sun and then rises up the chimney, turning turbines in the process. Fresh air is taken into the greenhouse and the cycle continues. The advantage of the solar chimney is that it does not require direct solar radiation and can operate on cloudy days even with solar radiation 100% diffused. The chimney generates 50 megawatts of electricity and is economical to operate.

SOLAR PONDS

A solar pond is a body of water in which the concentration of salt is so heavy that high temperatures can develop well

At the University of New Mexico, Professor Howard Bryant checks an experimental solar pond, which uses sunlight and the unique properties of salt water to generate enough power to heat an average-sized house.

below the surface. This discovery was made around 1900 by A. V. Kalecsinsky, who found that water at the bottom of several Hungarian lakes reached temperatures of 150°F in late summer. Kalecsinsky noted that solar radiation absorbed by these lakes resulted in an increase in temperature with the water's depth. In a freshwater pond, the water at the bottom warms up when it absorbs solar radiation. Then it expands and rises to the surface, where it cools off again. But because of the natural salt concentration in the Hungarian lakes, the lower regions remained denser even when warmer. This density prevented both thermal expansion and the convection currents that would have caused a transfer of heat out of the lake. Where no convection occurs, water can lose heat only by conduction—by direct contact with the air and the ground surrounding the water. But this heat-loss mechanism is slow, so that highly salty lakes get hot.

In natural solar ponds, the heavy salt concentration is maintained by salt deposits at the bottom of the ponds. The heavier the salt content of the water, the higher the temperature. The most efficient depth for a solar pond is 3 to 10 feet. The Great Salt Lake in Utah is a good example of this phenomenon. Under suitable conditions, it is possible to raise the temperature at the bottom of a solar pond even higher by increasing the salt content. It is, of course, also possible to create artificial solar ponds.

In addition to its relative simplicity, the solar pond can store a great deal of heat for long-term use. This ability contrasts dramatically with the heat storage capacity of the water tanks and rock beds used with more conventional systems. One solar pond might be able to heat several buildings of a housing complex. Small solar ponds are relatively easy to maintain. All that is needed is an infusion of a concentrated brine solution into the bottom of the pond from time to time. The top of the pond is "washed" with a weaker salt solution, and the overflow is removed. Thus, the salt carried away at the top of the pond compensates for the salt diffusing upward from the bottom, and the greater bottom concentration is maintained.

One problem encountered with solar ponds is keeping them clean, although it is still simpler than trying to clean vast areas of glass or mirrored surfaces. Particles of dirt on top of the pond can be washed away, and those that sink to the bottom cause no serious problem. However, dirt suspended within the water as a result of the high salt content can affect the amount of heat transmitted. Algae must be suppressed with chemicals.

Extensive experimentation with solar ponds has been going on near the Dead Sea in Israel since the early 1960s. These ponds were constructed with black bottoms to increase solar energy absorption. The ponds have reached temperatures of more than 194°F at a depth of 15 feet. However, each pond took two to three months to warm up after filling. By fitting large solar ponds with a specially designed turbine that only needs water at 194°F to produce electricity, it was possible to produce up to 5 megawatts of peak power with this technology. This experiment is only the beginning of a series of solar ponds for the Dead Sea that should produce 2,000 megawatts of electricity by the end of the century.

OCEAN THERMAL ENERGY

One of the more exotic methods for producing electricity from solar energy is ocean thermal energy conversion. Simply

Successor to the pedal-powered Gossamer Albatross, *the* Gossamer Penguin, *the world's first solar-powered airplane, cruises slowly over the Mojave Desert.*

put, this method uses differences in the temperature of different layers of ocean water to produce electricity.

In tropical oceans, surface water is often 30° to 35°F warmer than the water 1,500 feet below the surface. In 1881, Jacques Arsène d'Arsonval, a French physicist, found that a heat engine could be constructed using this temperature differential.

His basic design will be employed for a 390-foot, 25,000-ton power plant that will generate 100 megawatts of electricity, to be built on a floating platform off the coast of Indonesia by the middle of the 1990s. Surface water at 80°F will be pumped down to a boiler 200 feet below sea level. There the water's heat will be used to vaporize propylene, a liquid with a boiling point of just 70°F. As the pressurized gas rises through a closed loop of pipes, it will spin a dozen turbines, which will be hung in the water 30 feet above the boiler. From the turbines, the propylene gas will pass through a heat exchanger filled with water, at only 40°F, pumped up from as deep as 3,000 feet. The cold water will condense the propylene back into a liquid, which will then flow

In New Delhi, India, plastic sheeting has been used to construct solar stills that will produce clean drinking water.

back down to the boiler. The electricity produced will be sent to shore by means of underwater cables.

The basic concept of ocean thermal energy conversion is not new. First tested 60 years ago by a French engineer who built a pilot plant in Cuba, it has not been cost-effective. However, economic conditions today are such that it may be practical. Currently, Japan, Taiwan, and the United States are all funding ocean thermal energy research. Plans are on the drawing boards to place ocean thermal power plants off the coast of Florida in the warm waters of the Gulf Stream. There will be no need for energy storage—the ocean temperature remains fairly constant, and power can be generated even at night or on cloudy days. Although the potential of ocean thermal power potential is enormous, it is currently only in an experimental stage.

A technician checks the reflectivity of an individual heliostat. Keeping the surfaces of reflecting mirrors clean increases their efficiency.

chapter 7

THE PROS AND CONS OF SOLAR ENERGY

Many scientists believe that solar energy is the answer to the energy problems of the future. The most important advantage of solar energy is that it is virtually inexhaustible. The sun will continue to shine for billions of years, so its energy supply is practically infinite—no shortages to consider, no reserves to deplete, no wells to run dry.

Solar energy is also clean. It produces no air or water pollution and therefore generates no hidden costs for society to pay. A solar power plant may not be everyone's idea of an aesthetically pleasing structure, but there is no strip mining of land or flooding of canyons associated with it, and there is no fuel refining process that generates toxic by-products. There is no need to transport fuels that can spill and contaminate rivers and oceans. There is no problem of disposing of hazardous wastes.

Solar energy is not free, because of the expense of developing the technology to exploit it. But the energy source itself cannot be owned and monopolized, as fossil fuels have become because of the unequal geographic distribution of known reserves. Solar energy is almost equally available to all nations. Even though countries located in the equatorial regions will have more of it, no country will be so rich in energy resources that others without access to energy will be at its mercy.

There are disadvantages, too, however. Sunlight is diffuse and too weak to supply a large amount of energy at any one point. Although enough solar energy reaches the United States to supply the country's energy needs, it is spread very thinly and must be collected and concentrated. To use conventional solar collectors for this process on a large scale would take too much land. It is estimated that approximately 18,000 square miles would be needed to erect enough collectors to supply the total energy requirements of the United States. Back-up systems using other energy sources would be needed to supply solar heated homes and offices during cloudy periods or prolonged bad weather when stored energy might be depleted. It is unlikely that a total conversion to solar energy will be possible in the near future without some technological breakthrough.

One possible problem with the large-scale use of solar energy is the generation of thermal pollution, or excess heat. The whole point of solar collector systems is to turn light energy into heat energy. But thermal pollution can kill or injure marine life and may otherwise affect the ecology of an area. Excess heat, trapped in the earth's atmosphere by certain gases and pollutants, can contribute to global warming. Very high temperatures are reached in the solar collection process, but one way of handling this problem is to put the heat to work, converting it into other forms of mechanical energy or using the excess heat directly to heat offices and homes in large cities. Ultimately, however, because energy can be neither created nor destroyed, putting

A bizarre array of parabolic solar collectors alters the landscape of Warner Ranch, California. This system will provide electricity for 5,000 homes in the area.

heat energy to work moves it around and dissipates it but does not eliminate it or prevent its eventual accumulation.

THE COST OF SOLAR ENERGY

The biggest problem facing solar energy today is its cost. The initial cost of building solar collectors is considerably higher than that of building conventional power plants. In operation, the system would be very economical, since maintenance costs are minimal. The economic benefits increase with time, but the start-up costs are prohibitive.

Those who take the step to install solar energy systems may need to have special financial help. One way to encourage solar use is to offer income tax credits, such as those included in the energy program proposed by President Jimmy Carter in 1977.

The U.S. Department of Energy has made several proposals along these lines. It has suggested that aid be given to banks and other financial institutions that offer mortgages for new solar homes or for older homes retrofitted with solar equipment. It has also proposed incentives to insurance companies that provide coverage to such homes. Furthermore, the owners of rental dwellings could be encouraged to install solar systems through tax incentives. And subsidies could be provided to the construction industry to encourage them to design and build solar heated buildings. Currently, a number of states have enacted legislation designed to encourage solar energy development.

BARRIERS TO SOLAR ENERGY

One of the most formidable barriers to further development of solar energy is the language of many current building codes. Each American locality has its own code, the result of geography, climate, available building materials, and political forces. Of the 10,000 or more codes in this country, only a handful provide for the installation of solar collectors. If provisions for solar energy systems are not in the code, individuals are powerless to install them. Construction standards and safety rules must be worked out so that a system of solar collectors built by a homeowner could be approved by city building inspectors.

Another barrier is the lack of skilled labor for installing and maintaining solar energy systems. The financial incentives available for solar energy development in the 1970s dried up in the 1980s, and as a result, the construction industry has been slow to adapt to new technology. A solar engineer or architect with more than three years of experience is a rarity. The weakest link in the chain is the solar installation technician, who is often a plumber with no experience in solar installation problems. The lack of practical knowledge about the installation and operation of such systems often leads to poor machinery performance. General consumer dissatisfaction could have a crippling effect on the willingness of homeowners to take the risk of installing solar equipment.

Legal uncertainty is yet another obstacle. So far, only six states have dealt with the issue of the legal right to access sunlight. Light may have to pass through the air space of a neighbor's property before it is collected. Without legal protection, the po-

In front of the Sydney Opera House, crowds welcome the Quiet Achiever, *a solar-powered car that has just completed a 2,500-mile journey across Australia.*

In Boston, Massachusetts, onlookers examine the Solectria company's Lightspeed, a car powered by a combination of solar cells and electric storage batteries.

tential buyer of a solar collector system must think twice about committing money if his neighbor's whims, the growth of trees, or the construction of an apartment house block his access to sunlight.

Another stumbling block is the resistance of utility companies to solar energy. Since the collection of solar energy is difficult to control, utility companies are not anxious to see the technology expand. Some users of solar energy have been denied special lower-usage rates when the local utility company found out that they had a solar energy system. What is needed is a model solar heating and power code that forces utility companies to adopt progressive policies toward new clean-energy technologies.

SOLAR TRANSPORTATION

The development of photovoltaic cells as sources of direct electric power has tempted scientists and engineers to try new applications, most notably in the realm of transportation. Experimental cars and airplanes have been built, but so far they can only travel short distances. The total amount of solar energy that can be collected restricts the car's performance. It is like trying to drive a car at 50 miles per hour on the power of a hair dryer—roughly 1,500 watts. Such vehicles also have problems when the weather is cloudy.

All solar cars run on a trickle of electricity that is generated by solar cells on the top of the vehicles and stored in batteries that power an electric motor. The price of a typical solar panel capable of producing between 1,000 and 1,500 watts of electric power is about $15,000. A 5% increase in efficiency can be obtained by using high-quality silicon or gallium arsenide cells originally designed for satellites, but at a tenfold increase in price. The batteries used in these cars are of silver-zinc design, the lightest, most powerful available.

Unlike cars with gasoline engines that can weigh a ton or more, solar cars must be light. The ultimate construction material would be a blend of carbon fiber threads reinforcing a fiberglass frame. Forms would be joined with epoxy resin and cured for strength.

An artist's conception of the Powersat system proposed by Boeing. Large arrays of solar mirrors in earth's orbit will collect the sun's energy and beam it to the surface of the planet.

chapter 8

THE SOLAR FUTURE

In the past two decades, a great deal of attention has been focused on alternative sources of energy because of the political situation in the Middle East, where much oil originates. Even if the region were more stable politically, oil reserves would still be rapidly depleted. Oil is going to become very expensive some time within the next 30 years or so, and after that it may not be available at all. Extraction of oil from tar sands and the production of fuel oil from coal are complex and expensive processes that will only buy a short breathing space before fossil fuel reserves are completely exhausted. Furthermore, environmental concerns about air pollution, acid rain, oil spills, and global warming are also prompting a search for other energy sources.

New fuels such as alcohol made from grain are exciting interest, but the prospect of devoting large tracts of agricultural land to the production of motor fuel raises some serious questions about land degradation and the world's food supplies.

Nuclear power, once thought to be the ultimate solution to the energy crisis, is undergoing intense critical reevaluation. Escalating construction costs, the lack of a safe means of radioactive waste disposal, the potential for catastrophic accidents, and the danger of atomic fuel being converted into weapons all make the nuclear dream look more like a nightmare.

The sun can meet our energy needs indefinitely. Solar technology has been evolving for more than 2,000 years, but it took the oil embargo of 1973–74 to make people look seriously at solar energy's potential. In less than 20 years, a revolution has taken place in solar energy applications. The idea of building huge solar power stations to capture energy from the sun used to belong to the world of science fiction, but no longer.

Israel now leads the world in the practical application of solar power. A third of all Israeli homes get their hot water from solar collectors. Japan has embarked on an ambitious solar energy program. Since it must import all but 2% of its required fuel, the use of solar energy is encouraged. The Japanese government is financing the development of a large-scale centralized electric generating facility powered by solar energy. In addition, solar water heaters are being mass-produced in Japan and even exported to other countries.

The French government has established the Institute of Solar Energy to coordinate the work of more than 300 scientists and engineers working in different areas of solar research. The Swiss are considering installing solar heating and cooling systems throughout their country, and the Germans are committed to a multimillion-dollar solar research program.

The Soviet Union is concentrating on the development of solar power systems for its undeveloped regions, such as Siberia. Soviet scientists have already designed solar refrigerators, solar powered sluice gates for irrigation projects, and solar stills to provide sheep and cattle with drinkable water in desert areas.

In rural China, some peasants are experimenting with solar furnaces to cook their meals.

Australia has made the most serious commitment to solar energy of any country in the world. In certain parts of the country, solar hot water heaters are required by law, as are solar stills in areas without drinkable water.

THE UNITED STATES

Of all the countries in the world, the United States is the best able to afford solar energy. With 6% of the world's population, it uses 33% of the world's energy, and its rate of use has been doubling every 15 years. The introduction of solar energy into American life must be a planned venture between the government and free enterprise.

In 1974, the U.S. Congress took the first step in this direction by passing the Solar Energy Research Development and Demonstration Act. The purpose of this act was to pursue a vigorous program of research and resource assessment. The act established the Solar Energy Research Institute near Golden, Colorado, which is responsible for assessing all solar energy technologies, with the goal of increasing the efficiency of these

This new, high-efficiency parabolic solar collector, designed by the McDonnell Douglas Corporation, can generate 25 kilowatts of electricity.

technologies and bringing costs down to a point where solar energy is an economical energy alternative. The act also seeks to foster the creation of a viable commercial solar energy industry.

The U.S. Department of Energy has divided solar energy technologies into three major groups: (1) thermal applications (heating and cooling of buildings); (2) solar electric power (photovoltaic cells, solar power towers, and ocean thermal energy); and (3) fuels from biomass (the burning of plant matter, including wood and waste products).

BIOMASS CONVERSION

Biomass conversion is a new technology that uses plant materials to produce electric power. Green plants convert the energy of the sun into food and plant tissue during the process of photosynthesis. Over a period of millions of years, the stored energy of these plants is transformed into fossil fuels and used as a source of power.

Biomass conversion speeds up the process by which the energy stored in plants is converted into fuel. One method involves the growing of large amounts of plant material to be used specifically as fuel. Plants such as eucalyptus, sugarcane, and water hyacinth grow quickly and provide an abundant source of energy. Scientists envision large "energy plantations" where such plants could be grown and then burned to provide heat and steam to drive electric turbines. Other schemes would treat biomass materials with chemicals, converting them into methane and other gases that could be used as fuels, thus avoiding the release of large amounts of carbon into the air. Seaweed kelp is well suited to this method.

SOLAR HYDROGEN

Electricity produced from the collection of solar energy can be used to produce hydrogen by splitting water into its component elements through electrolysis. Hydrogen is a clean fuel that can be used for transportation and heating. When it undergoes combustion with oxygen, it simply becomes water again. The process produces neither carbon monoxide, carbon dioxide, hydrocarbons, nor particulate matter. The only pollutants produced are oxides of nitrogen, which can be reduced to very low levels. Hydrogen also can be used to generate electricity, like any other fuel, by creating heat and using the heat to generate

Inventor Charles Eames demonstrates his solar-powered Thingamajig, which spins its wheels and creates strange optical and sound effects using only the energy of the sun.

Unequal heating of the earth's surface by the sun produces regional differences in atmospheric pressure, which create winds. Wind is therefore considered a form of solar energy, and it is rapidly being exploited to produce electricity at wind farms such as this one in California.

steam. And it can be used with exotic new devices, such as the highly efficient fuel cells used in spacecraft. Sunny deserts are prime sites for hydrogen production, because the water requirements for electrolysis are modest, equal to only two to three centimeters of rainfall per year.

NEW WAYS OF THINKING

Solar technology requires new ways of thinking about energy. The energy compacted into fossil fuels is so enormous

Part of the solar research facility at Odeillo, France, this huge solar generator is the world's most powerful, with an output of 1,000 kilowatts.

that industrial civilization has used it wastefully. The collection of solar energy, however, is an expensive and technologically difficult process that demands more efficient use of what is collected. The use of solar energy in small-scale units for residential heating and cooling requires homeowners to be more conservation-minded in their use of energy so as to reap its maximum benefits. Large-scale commercial use requires big investments in expensive solar hardware, and the users of such energy must accept that it will not be as cheap as fossil fuels. The widespread introduction of solar energy systems also will have a profound effect on our economy. The production of energy will be put in the hands of many consumers, and the utility companies will be unable to control the supply of this energy source. Only a large corporation can raise the funds to drill for oil or build refineries. But many small companies can provide the hardware for solar collectors. Whereas urban centers will still be dependent on corporate power grids, whether the source of energy is a fossil fuel power plant or a large array of mirrors, many homeowners and small businesses away from the major cities will have more options. Utility companies are not likely to surrender to new economic order without a struggle. They are likely to fight for expansion of the types of energy production that they can control—nuclear and hydroelectric power, for example.

Despite dramatic progress in recent decades, solar energy is still not economically competitive with conventional energy generation, at least in terms of costs to the firm, if not in terms of social or environmental costs. Because of this impediment, more cooperation is necessary among researchers, producers, and potential users so that solar energy technologies can be developed rapidly.

A major challenge will be the creation of a viable economic and industrial system for the exploitation of solar energy. This system should include favorable tax policies and incentives for solar research and equipment manufacturing. Whether society chooses to invest in this technology or not, the era of fossil fuels is coming to an end, and some alternative source of energy will have to be found in the near future.

APPENDIX: FOR MORE INFORMATION

Environmental Organizations

American Institute of Architects
1735 New York Avenue NW
Washington, DC 20006
(202) 626-7300

American Museum of Science
  and Energy
300 South Tulane Avenue
Oak Ridge, TN 37830
(615) 576-3218

American Solar Energy Society
2400 Central Avenue G-1
Boulder, CO 80301
(303) 443-3130

Conservation and Renewable
  Energy Inquiry and Referral
  Service
(800) 523-2929

Council for Renewable Energy
  Education
777 North Capitol Street NE
  Suite 805
Washington, DC 20002
(202) 408-0309

Environmental Action Foundation
6930 Carroll Avenue, Suite 600
Takoma Park, MD 20912

Fund for Renewable Energy and
  the Environment
1001 Connecticut Avenue NW
  Suite 638
Washington, DC 20036
(202) 466-6880

National Energy Information
  Center
EI-231, 1000 Independence
  Avenue SW
Washington, DC 20585
(202) 586-8800

New Alchemy Institute
237 Hatchville Road
East Falmouth, MA 02536
(508) 564-6301

Renew America
1001 Connecticut Avenue NW
  Suite 719
Washington, DC 20036
(202) 232-2252

Resources for the Future
1616 P Street NW
Washington, DC 20036
(202) 328-5000

Solar Energy Research Institute
1617 Cole Blvd.
Golden, CO 80401
(303) 231-1378

Government Agencies

Department of Energy
1000 Independence Avenue SW
Washington, DC 20585
(202) 586-5000

Conversion Table

(From U.S./English system units to metric system units)

Length

1 inch = 2.54 centimeters
1 foot = 0.305 meters
1 yard = 0.91 meters
1 statute mile = 1.6 kilometers (km.)

Area

1 square yard = 0.84 square meters
1 acre = 0.405 hectares
1 square mile = 2.59 square km.

Liquid Measure

1 fluid ounce = 0.03 liters
1 pint (U.S.) = 0.47 liters
1 quart (U.S.) = 0.95 liters
1 gallon (U.S.) = 3.78 liters

Weight and Mass

1 ounce = 28.35 grams
1 pound = 0.45 kilograms
1 ton = 0.91 metric tons

Temperature

1 degree Fahrenheit = 0.56 degrees Celsius or centigrade, but to convert from actual Fahrenheit scale measurements to Celsius, subtract 32 from the Fahrenheit reading, multiply the result by 5, and then divide by 9. For example, to convert 212° F to Celsius:

212 − 32 = 180 x 5 = 900 ÷ 9 = 100° C

FURTHER READING

Ametek, Inc., Power Systems Group. *Solar Energy Handbook—Theory and Applications.* Radnor, PA: Chilton, 1979.

Anderson, Bruce, and Michael Riordan. *The New Solar Home Book.* Andover, MA: Brick House, 1987.

Buckley, Shawn. *Sun Up to Sun Down.* New York: McGraw-Hill, 1979.

Cassidy, Bruce. *The Complete Solar House.* New York: Dodd, Mead, 1977.

Coe, Gigi. *Present Value—Constructing a Sustainable Future.* San Francisco: Friends of the Earth, 1979.

Crowther, Richard. *Sun/Earth, How to Use Solar and Climatic Energies.* New York: Scribners, 1977.

Daniels, George E. *Solar Homes and Sun Heating.* New York: HarperCollins, 1976.

Green, Martin A. *Solar Cells: Operating Principles, Technology, and Application.* Englewood Cliffs, NJ: Prentice-Hall, 1982.

Greenwald, Martin L., and Thomas K. McHugh. *Practical Solar Energy Technology.* Englewood Cliffs, NJ: Prentice-Hall, 1985.

Hayes, Dennis. *Rays of Hope: The Transition to a Post-Petroleum World.* New York: Norton, 1977.

Kendall, Henry W., and Steven Nadis. *Energy Strategies: Toward a Solar Future.* Cambridge, MA: Ballinger, 1980.

Kreider, Jan F. *Handbook of Solar Energy*. New York: McGraw-Hill, 1980.

Schepp, Brad. *The Complete Passive Solar Home Book*. Blue Ridge Summit, PA: Tab Books, 1985.

Solar Magazine Editors. *Solar Age Resource Book*. New York: Everest House, 1979.

Starr, Gary. *The Solar Electric Book*. Lower Lake, CA: Integral, 1987.

Weinberg, Carl J., and Robert H. Williams. "Energy from the Sun." *Scientific American* (September 1990): 146–55.

Yokell, Michael P. *Environmental Benefits and Costs of Solar Energy*. Lexington, MA: Lexington Books, 1980.

GLOSSARY

active solar energy system   A mechanical system that uses solar collectors to heat and cool a building.

array   A group of solar mirror panels covered with **photovoltaic cells** that collect energy from the sun.

atmosphere   The layer of gases that surrounds a planet, such as the air that surrounds the earth.

bioconversion   The process of using solar energy stored in green plants by burning them to produce electrical power.

biomass   Organic materials—such as wood, crops, and other material—that are used as fuel during **bioconversion**; also refers to the total mass of living matter in a given area.

chromosphere   A red-colored layer of hot gases that makes up the outer shell of the sun's **atmosphere.**

conductor   Any material capable of transmitting electricity.

convective zone   The outer part of the sun where currents of hot gas rise and sink, helping the sun's energy to escape.

electrolysis   A process in which an electric current is passed through a liquid, causing a chemical reaction to take place; in this process, water is separated into its two elements, hydrogen and oxygen.

electron   A small, negatively charged particle that orbits the nucleus of an atom.

flat-plate solar collector   A collector consisting of a flat metal plate painted black and covered with a piece of glass or plastic; water or air circulates through pipes attached to the back of the plate and picks up heat absorbed by the collector.

focusing solar collector   A collector that uses a system of curved mirrors to focus the sun's rays on a central point.

fossil fuels   Combustible materials, including coal, oil, and natural gas; formed over millions of years, under conditions of high temperature and pressure, from the remains of plants and animals.

generator   A machine that converts mechanical energy into electrical energy.

greenhouse effect   Solar heating using an enclosure of glass or other transparent materials that admit **solar radiation** but block the passage of heat energy radiated back from a surface that has absorbed sunlight; also refers to the trapping of infrared radiation in the earth's atmosphere that results in global warming.

heliostat   An individual solar collecting mirror.

insolation   The amount of solar energy that strikes an object.

insulation   Construction materials that can stop or slow down the rate at which heat passes through them.

ocean thermal energy conversion   The process of using the temperature differential between the warm water at an ocean's surface and the much colder water below to produce electric power.

parabolic trough   A device used to collect solar energy in which a parabolic mirror focuses the sun's rays on a point.

photosphere An outer layer of the sun that consists of dense, hot gases that allow light to escape; the part of the sun that can be seen from the earth.

photovoltaic cell A solar energy device that converts sunlight directly into electricity; composed of the p-type layer, which contains positive charges that attract excess electrons, and the n-type layer, which provides the excess electrons.

radiant energy Energy that is emitted in waves, or radiated, from a particular place such as a fire or the sun.

radiative zone The inner part of the sun, where energy from the core radiates outward to the convective layers.

retrofitting Erecting a solar collection system on an existing building to make it more energy efficient.

solar collector A device used to collect and concentrate the rays of the sun; contains a reflective surface as well as a medium for conducting the energy collected.

solar furnace A focusing collector used to create very high temperatures by means of a number of mirrors concentrating the sun's rays on a small point.

solar panel A large, flat object that can easily absorb heat from the sun.

solar pond An artificial body of water containing a stratified salt solution; stores solar energy as heat, because the presence of the salt solution reduces the loss of heat from the pond.

solar radiation   Heat and light energy that travels through space in the form of electromagnetic waves.

solar still   A device that separates water from impurities, such as salt, through the processes of evaporation and condensation.

solar thermal conversion   The process of using solar energy to generate electricity by means of steam-driven electric turbines; the steam is produced from water heated by solar radiation.

solar tower   A solar energy collecting structure that is positioned to collect reflected solar radiation from an array of **heliostats**; the top of the tower contains the heat exchange chamber, and a hot fluid is used to power an electrical generating system at ground level; also known as a power tower.

INDEX

Abbot, Charles, 28–29
Acid rain, 20
Adams, W., 28
Africa, 57
Apep, 25
Archimedes, 26
Arco Solar, 53
Arrays, 51, 60
Atacama Desert, 70
Auroras, 16
Australia, 35, 39, 57, 89

Bailey, William J., 32
Baltimore, Maryland, 31
Barstow, California, 44
Belgium, 55
Bell Laboratories, 54
Biomass conversion, 91
Bombay, India, 28
Boyle, John, Jr., 33
British Isles, 55
Buffon, George, 28

Cairo, Egypt, 33
Caldarium, 26
Canada, 39
Carbon, 14
Carbon dioxide, 20, 92
Carbon monoxide, 92
Carinna Plain solar power station, 53
Carter, Jimmy, 81
Ceramics, 17

Chichén Itzá, 24
Chromosphere, 17
Climax, 31
Coal mining, 20
Coconut Island, 57
Conduction, 16–17, 59
Convection, 16, 17, 59
Coober Pedy, Australia, 70
Corona, 17
Coronagraph, 18
Cuba, 77

Dark Ages, 27
D'Arsonval, Jacques Arséne, 75–76
Day and Night, 31–32
Dead Sea, 74
De Architectura (Vitruvius), 26
Department of Energy, U.S., 44, 82, 91
De Saussure, Nicolas, 28
Dositheius, 25

Einstein, Albert, 14, 50
Einstein's equation, 14
Electric Power Research Institute, 55
El Paso, Texas, 45
Eneas, Aubrey C., 33
Energy, 13. *See also* Fossil fuel; Hydroelectric energy; Nuclear energy; Solar energy
England, 24
Ericsson, John, 29, 30–31

Eucalyptus, 91
European Community (EC), 55
European Phoebus project, 44

Fossil fuel, 19, 20, 49, 56, 87, 93, 95
France, 29, 55, 88
Fritts, Charles, 53
Fusion, 15

Germany, 55, 88
Glass, 17
Golden, Colorado, 90
Great Salt Lake, 73
Greenhouse effect, 20
Gulf Stream, 77

Harding, J., 34
Harper Lake, California, 46
Heliocaminus, 26
Helium, 14, 15
Hottinger, M. M., 35
Hydroelectric energy, 19
Hydrogen, 14, 15, 92–93
Hydrogen bomb, 15

Insolation, 38, 52
Institute of Solar Energy, 88
International Solar Energy Society, 35
Israel, 39, 74, 88
Italy, 55

Japan, 35, 39, 40, 52, 77, 88

Kalecsinsky, A. V., 73
Kemp, Clarence M., 31
Kilowatt, 13

Las Salinas, Chile, 70
Lavoisier, Antoine, 28
Leonardo da Vinci, 27
Leslie Salt Company, 70
Los Angeles, California, 46
LUZ Corporation, 46

Magnesium, 14
Manzanares, Spain, 71
Middle Ages, 31
Mojave Desert, 44, 46
Monitor, 30
Monte Alban, 24
Mouchot, Augustin, 29–30

Natural Bridges National Monument, 54
Needles, California, 33
New Guinea, 57
Nio, Japan, 40
Nitrogen, 14
Nuclear energy, 19, 87

Odeillo, France, 28, 40, 44
Oxygen, 14

Paraffin, 60
Pasadena, California, 33
Pellworm, 55
Photosphere, 17
Photosynthesis, 19
Photovoltaic cells, 49–50, 51, 52–54, 55, 56, 57, 60, 85
Pifre, Abel, 30
Planck, Max, 50
Pliny the Younger, 26
Power, 13
Proxima Centuri, 14

Pyrenees Mountains, 28

Quanta, 50

Ra, 24–25
Radiation, 16, 17, 59

San Francisco, California, 70
San Luis Obispo, California, 53
Scientific American, 28
Seaweed kelp, 91
Semiconductors, 50–51
Shuman, Frank, 33–34
Siberia, 88
Silicon, 14, 85
Solar cells. See *Photovoltaic cells*
Solar chimney, 71–72
Solar cooker, 70–71
Solar distillation, 69–70
Solar eclipse, 17–18, 23–24
Solar energy
    active, 59–60
    advantages, 19–20, 79-80
    and agriculture, 23, 56
    collectors for space heating, 37–47
    construction of power stations, 82-83
    cost, 81–82
    drawbacks, 20–21, 80
    early experiments, 28–31
    inventions that heat water, 31-35
    legalities, 83–84
    and medical care, 56
    opposition by utility companies, 82-85, 95
    passive, 59

    storage, 60–63
    systems, 59
    in transportation, 85
    world technology, 88
Solar Energy Research Development and Demonstration Act, 90–91
Solar Energy Research Institute, 90
Solar One power plant, 44
Solar pond, 72–74
Solar power tower, 41, 42–45
Soviet Union, 88
Spain, 55
Stonehenge, 24
Strip-mining, 20
Sugarcane, 91
Sun
    and ancient cultures, 24–27, 31, 64
    dimensions, 14
    elements, 14
    as energy source, 15–16, 17, 18–19, 20
    gaseous layers, 14–15, 17–18
    and religious beliefs, 23
Switzerland, 88
Syracuse, 26

Taiwan, 77
Telescope, 18
Torres Strait, 57
Tuat, 25

United States, 19, 20–21, 29, 35, 40, 54, 55, 67, 77, 80, 89–91

Vanguard satellite, 54

Vitruvius, 26

Water hyacinth, 91
Watt, 13
Willsie, H. E., 33
Wilson, C., 34
Wind energy, 19, 21

Wood, 17
World War I, 34
World War II, 34

Yucatán Peninsula, 24

Zurich Institute of Technology, 35

PICTURE CREDITS

AP/Wide World Photos: pp. 12, 16, 18, 22, 32, 36, 42 (top), 43, 48, 51, 56, 58, 78, 81, 84, 86, 89, 94; Australian Information Service, photograph by Peter Kelly, courtesy of the Australian Consulate: p. 83; Australian Information Service, photograph by Bob Peisley, courtesy of the Australian Consulate: p. 21; The Bettmann Archive: pp. 27, 29; Photo courtesy of Harry Braun/Research Analysts, Mesa, AZ: p. 90; Reuters/Bettmann Archive: pp. 68, 71; Solar Energy Research Institute, photo courtesy of U.S. Windpower: p. 93; Steve Strong and Associates: p. 66; UN Photo 151,416/J. K. Isaac: p. 76; UN Photo 150768/Derek Lovejoy: p. 34; UPI/Bettmann Archive: pp. 20, 30, 42 (bottom), 46, 47, 61, 63, 65, 72, 75, 92

ABOUT THE AUTHOR

BOB BROOKE is a public school teacher and an author who has written more than 1,300 articles over the past 17 years in the fields of business, education, travel, art, gardening, and the environment. He is the author of *The Amish Country*, a portrait of the Amish people; *Garden by Recipe*, a guide to home gardening; and *Best Places to Stay in the U.S.A.*, an accommodations guide to the United States.

ABOUT THE EDITOR

RUSSELL E. TRAIN, currently chairman of the board of directors of the World Wildlife Fund and The Conservation Foundation, has had a long and distinguished career of government service under three presidents. In 1957 President Eisenhower appointed him a judge of the United States Tax Court. He served Lyndon Johnson on the National Water Commission. Under Richard Nixon he became under secretary of the Interior and, in 1970, first chairman of the Council on Environmental Quality. From 1973 to 1977 he served as administrator of the Environmental Protection Agency. Train is also a trustee or director of the African Wildlife Foundation; the Alliance to Save Energy; the American Conservation Association; Citizens for Ocean Law; Clean Sites, Inc.; the Elizabeth Haub Foundation; the King Mahendra Trust for Nature Conservation (Nepal); Resources for the Future; the Rockefeller Brothers Fund; the Scientists' Institute for Public Information; the World Resources Institute; and Union Carbide and Applied Energy Services, Inc. Train is a graduate of Princeton and Columbia Universities, a veteran of World War II, and currently resides in the District of Columbia.